VCE Units 3 & 4
PSYCHOLOGY

PETER MILESI

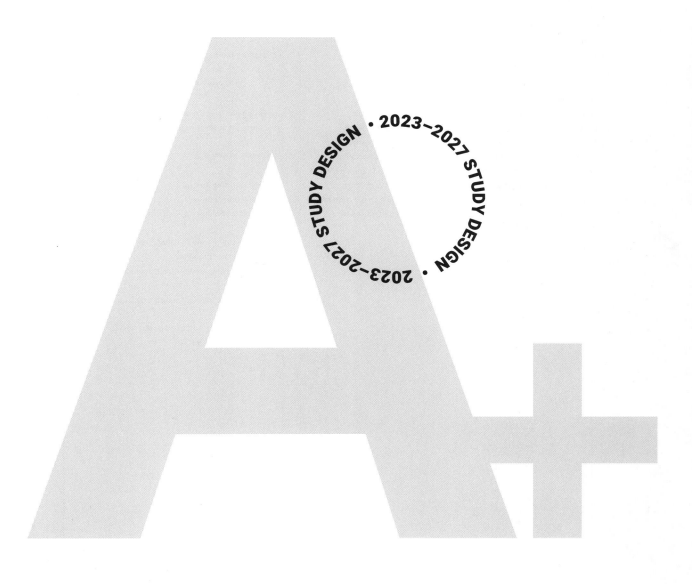

- + summary notes
- + revision questions
- + detailed, annotated answers
- + study and exam preparation advice

STUDY NOTES

A+ VCE Units 3 & 4 Psychology Study Notes
1st Edition
Peter Milesi
ISBN 9780170465212

Publisher: Alice Wilson
Series editor: Catherine Greenwood
Copyeditor: Leanne Peters
Reviewer: Amelia Brear
Consultant: Meredith McKague
Series text design: Nikita Bansal
Series cover design: Nikita Bansal
Series designer: Cengage Creative Studio
Artwork: MPS Limited
Production controller: Karen Young
Typeset by: Nikki M Group Pty Ltd

Any URLs contained in this publication were checked for currency during the production process. Note, however, that the publisher cannot vouch for the ongoing currency of URLs.

Acknowledgements
Selected VCE examination questions and extracts from the VCE Study Designs are copyright Victorian Curriculum and Assessment Authority (VCAA), reproduced by permission. VCE ® is a registered trademark of the VCAA. The VCAA does not endorse this product and makes no warranties regarding the correctness or accuracy of this study resource. To the extent permitted by law, the VCAA excludes all liability for any loss or damage suffered or incurred as a result of accessing, using or relying on the content. Current VCE Study Designs, past VCE exams and related content can be accessed directly at www.vcaa.vic.edu.au.

© 2023 Cengage Learning Australia Pty Limited

Copyright Notice
This Work is copyright. No part of this Work may be reproduced, stored in a retrieval system, or transmitted in any form or by any means without prior written permission of the Publisher. Except as permitted under the *Copyright Act 1968,* for example any fair dealing for the purposes of private study, research, criticism or review, subject to certain limitations. These limitations include: Restricting the copying to a maximum of one chapter or 10% of this book, whichever is greater; providing an appropriate notice and warning with the copies of the Work disseminated; taking all reasonable steps to limit access to these copies to people authorised to receive these copies; ensuring you hold the appropriate Licences issued by the
Copyright Agency Limited ("CAL"), supply a remuneration notice to CAL and pay any required fees. For details of CAL licences and remuneration notices please contact CAL at Level 11, 66 Goulburn Street, Sydney NSW 2000,
Tel: (02) 9394 7600, Fax: (02) 9394 7601
Email: info@copyright.com.au
Website: www.copyright.com.au

For product information and technology assistance,
in Australia call **1300 790 853**;
in New Zealand call **0800 449 725**

For permission to use material from this text or product, please email
aust.permissions@cengage.com

ISBN 978 0 17 046521 2

Cengage Learning Australia
Level 7, 80 Dorcas Street
South Melbourne, Victoria Australia 3205

Cengage Learning New Zealand
Unit 4B Rosedale Office Park
331 Rosedale Road, Albany, North Shore 0632, NZ

For learning solutions, visit **cengage.com.au**

Printed in Singapore by C.O.S. Printers Pte Ltd.
1 2 3 4 5 6 7 27 26 25 24 23

CONTENTS

PREPARING FOR THE END-OF-YEAR EXAM vi
COMMON COMMAND TERMS USED IN ASSESSMENT xi
HOW TO USE THIS BOOK xii
A+ DIGITAL FLASHCARDS xiv
ABOUT THE AUTHOR xiv

UNIT 3 HOW DOES EXPERIENCE AFFECT BEHAVIOUR AND MENTAL PROCESSES?

Chapter 1 Area of Study 1: How does the nervous system enable psychological functioning?

Area of Study summary		2
Summary of Units 1 & 2 prior knowledge		4
1.1	Nervous system functioning	4
	1.1.1 Divisions of the nervous system	4
	1.1.2 Role of neurotransmitters in transmission of neural information across a neural synapse	7
	1.1.3 Role of neuromodulators and their effect on brain activity	10
	1.1.4 Synaptic plasticity	12
1.2	Stress as an example of a psychobiological process	15
	1.2.1 Stress response	16
	1.2.2 Gut–brain axis	18
	1.2.3 General Adaptation Syndrome	19
	1.2.4 Transactional Model of Stress and Coping	21
	1.2.5 Coping strategies	23
Glossary		26
Revision summary		31
Exam practice		35

Chapter 2 Area of Study 2: How do people learn and remember?

Area of Study summary		48
2.1	Models to explain learning	50
	2.1.1 Approaches to understand learning	50
	2.1.2 Classical (respondent or Pavlovian) conditioning	50
	2.1.3 Operant (instrumental) conditioning	54
	2.1.4 Observational learning (modelling)	59
	2.1.5 Approaches that situate the learner within a system	61
	2.1.6 Comparison of learning theories	62
2.2	The psychobiological process of memory	64
	2.2.1 Background knowledge underpinning models for explaining memory	64
	2.2.2 Multi-store model of memory	65
	2.2.3 Key brain regions involved in memory	71
	2.2.4 Role of episodic and semantic memory	72
	2.2.5 Use of mnemonic devices	74
Glossary		77
Revision summary		81
Exam practice		86

UNIT 4 HOW IS MENTAL WELLBEING SUPPORTED AND MAINTAINED?

Chapter 3 Area of Study 1: How does sleep affect mental processes and behaviour?

Area of Study summary	100
Background knowledge	102
3.1 The demand for sleep	105
3.1.1 Measurement of responses that can indicate different stages of sleep	105
3.1.2 Sleep	109
3.1.3 The biological mechanisms of sleep	112
3.1.4 Sleep patterns across the life span	114
3.2 Importance of sleep to mental wellbeing	114
3.2.1 Sleep deprivation	114
3.2.2 Circadian rhythm sleep disorders	116
3.2.3 Improving sleep–wake patterns	118
Glossary	121
Revision summary	124
Exam practice	127

Chapter 4 Area of Study 2: What influences mental wellbeing?

Area of Study summary	139
Summary of Units 1 & 2 prior knowledge	141
4.1 Defining mental wellbeing	146
4.1.1 Mental wellbeing	146
4.1.2 Social and emotional wellbeing	146
4.1.3 Mental health continuum	147
4.2 Application of a biopsychosocial approach to explain specific phobia	150
4.2.1 Specific phobia	150
4.3 Maintenance of mental wellbeing	157
4.3.1 Protective factors that maintain mental wellbeing	157
Glossary	161
Revision summary	164
Exam practice	168

Chapter 5 Area of Study 3: How is scientific inquiry used to investigate mental processes and psychological functioning?

Area of Study summary		178
5.1	The research process	179
	5.1.1 Operational hypotheses	180
	5.1.2 Methodologies used in psychological research	180
	5.1.3 Ways to minimise the effects of extraneous variables	182
	5.1.4 Participant selection	183
	5.1.5 Participant allocation	184
	5.1.6 Placebo effect	184
	5.1.7 Experimenter effect	185
	5.1.8 An overview of three experimental designs	185
	5.1.9 Statistics	186
	5.1.10 Reliability and validity of results	189
	5.1.11 Appropriateness of conclusions and generalisations based on results obtained	190
	5.1.12 An overview of ethical considerations in the conduct of psychological research	190
	5.1.13 Use of non-human animals in research	193
	5.1.14 American Psychological Association format for reporting psychological research	194
5.2	Student-designed practical investigation	196
	5.2.1 Logbooks	196
	5.2.2 Scientific posters	196
Glossary		200
Revision summary		204
Exam practice		210

ANSWERS 217

PREPARING FOR THE END-OF-YEAR EXAM

Getting ready for your exams does *not* begin the night before, or during Swot Vac, or in any of the various revision programs that occur in the month before the exam.

To achieve your optimal results for the examination, you should begin your preparation as soon as possible, ideally from the beginning of the year.

Effective preparation involves the following steps.

1 During the year

Be organised!

Keep up to date with your notes and your class work. If you fall behind in your work, later concepts may not make sense.

Organise your study routine

Apply effective study habits and techniques over a period of time. You cannot leave things until the last minute, nor can you expect to take in all of the content during the week before your examination if you have not laid the foundations during the year.

Whether you have homework or not, you should spend some time every night going over your notes or reading your text. Regular revision and relearning helps to encode information into your memory and reinforce semantic networks organised according to meaning and based on your understanding of concepts.

When revising, work actively (e.g. underlining, highlighting, jotting down points, drawing diagrams, graphs, flow charts and concept maps and doing many sample questions). You will remember what you read much better if you use it in some way. By dealing with information, you have to think about it, which helps reinforce links to your semantic networks and prevents you from falling into a daze and daydreaming.

Test yourself regularly. This not only gives you a regular check on your understanding of the material, it also gives you practice in the skills required to answer questions for the exam.

When studying, avoid all other distractions so that your attention is focused. This will not only help information to be absorbed and encoded into your memory but will also make the process faster and more efficient.

Use your class time wisely

Become involved in class discussions and activities, especially those applying concepts to a practical context.

Ask questions (to peers as well as your teacher) whenever you don't understand something.

Keep clear notes

Take detailed notes *in your own words* and include as many original examples as you can. This will help to reinforce your understanding of the concepts.

Keep a glossary

This should not just include key terms or theories from the Study Design, but any terms you feel need clarification. Where appropriate, try to include an example with your definition to reinforce your understanding.

2 Leading up to the exam

Read and research

Read your text *and* as many other sources as you can. Wider reading not only gives you more information about the concepts and theories, but allows you to see information in a different form, which may also help to clarify points and assist in your understanding of the information.

Be physically and mentally prepared

Make sure you have adequate sleep, relaxation and exercise, as well as a balanced diet. These things should be incorporated into your study routine along with your work. Also, know when to give yourself a break and when to stop studying. To be able to maintain your efficiency, you must take regular breaks as part of your study routine.

Develop a positive attitude about your abilities. Your marks will reflect the amount of work and preparation you have put in to enhance your potential, but you can only aim to do your best on the day of the exam.

Prepare carefully

Plan a revision timetable in advance that will eventually build up to the exam itself – then stick to it!

Continue to work actively and test yourself regularly on small blocks of work. Read a chapter or section of your notes, then close your book and try to recall as much as possible.

Develop mnemonics

Mnemonics are techniques to enhance your ability to recall information. Although you can learn these strategies from others, you will probably find that ones you invent yourself are easier to remember and apply. It does not matter if some of the images or associations seem absurd or ridiculous – if they help you to remember and understand material for the exam, then they have served their purpose.

Organise study groups

Working in groups can enhance your revision. By discussing concepts with your peers, you are reinforcing your understanding. When you can explain something to others in the group, then you will know that you understand that concept. Conversely, others in the group may be able to help you with concepts that you do not know well, especially as things may 'click' when you hear something explained in a different way from how your teacher or textbook explains it.

Study groups can also provide you with some much-needed moral support if you start to get anxious about the exam and your ability to perform.

Do trial exams and/or past papers

These will not only show you the type of questions to expect, but will also give you practice in writing answers. It is a good idea to make the trial exams as much like the real exam as possible (in terms of conditions, time constraints, materials etc.).

Use trial exams and comments from your teacher to pinpoint weaknesses and work to improve these areas. While you may concentrate on what you don't know well, also remember to revise all of the other areas.

Make sure that you understand the material rather than try to rote learn definitions, especially as much of the Psychology examination will aim to assess how well you can apply the information within the course.

Check your state of readiness

- Do you know the time and date of the exam?
- Have you prepared summaries for all of the areas of study? Do you read these regularly?
- Do you understand all the concepts within the course?
- Have you looked at previous exam papers?
- Have you practised writing exam answers?
- Have you done a trial exam under conditions that approximate the actual exam?

3 The night before the exam

Gather together everything you will need for the exam
This includes HB pencils, a pencil sharpener, an eraser, pens and a ruler. Once collected, put them somewhere appropriate so that you will not leave them behind on the morning of the exam.

Skim lightly over your notes
Revise normally. Don't try last-minute cramming.

Go to bed early
You may find it hard to get to sleep if you are anxious about the exam. Being rested will help you to concentrate better during the exam, and achieve your best possible results.

4 On the day of the exam

Don't try to learn anything new
This may interfere with what you have already learned, leading to confusion and increased stress levels.

Get to school/the exam venue with time to spare
If you are rushing to the exam, you will only increase your stress levels and reduce your ability to do well. Furthermore, if you are late starting the exam, you may not have time to answer all of the questions or you might have to rush to finish (and so increase your chance of making careless mistakes). If you are too late, you will *not* be allowed to enter the exam.

Don't worry too much about 'exam jitters'
A certain amount of stress is required to help you concentrate and achieve an optimum level of performance. However, if you're still feeling very nervous, try some relaxation techniques, such as breathing exercises, to calm yourself down.

Don't discuss things too much with your peers
This will probably only increase your stress levels, especially if someone asks you a question to which you do not know the answer.

5 During the exam

The format of the exam
The examination for VCE Psychology will be held at the end of the year and will only cover material in the Study Design for Units 3 & 4. It is worth 50% of your overall study score.

The exam has 15 minutes' reading time and 150 minutes' (2.5 hours) writing time and is worth 140 marks (effectively just over one minute per mark). The exam is in two sections (multiple choice and short answer, including extended response), and each section of the exam covers all areas of the course content, with some questions potentially covering more than one area of study within them.

Reading time

Use your reading time wisely
You have 15 minutes to read the exam paper. Read the instructions carefully, noting any directions that must be carried out. *Do not use reading time to try to figure out the answers to any of the questions until you have read the whole examination!* Read all of the questions to get an overall picture of the examination.

One approach is to read the short-answer questions first, as some of the material in the multiple-choice section may trigger recall or stimulate thinking for your answers.

Look for key words in the question

To understand what the question is asking, analyse it for key words or instructions. Highlight or underline these later, before you attempt the question.

Plan your approach to the examination

Leave the most difficult questions until last and go back to them when you have answered the rest. Beginning with questions that you find easy will boost your confidence and give you momentum to finish the exam within the prescribed time.

As some responses may come to mind during reading time, many students prefer to begin with the short-answer questions while answers are still fresh in their mind. Also, as short-answer questions tend to take longer, doing them first gives you the option of quickly putting an answer to multiple-choice questions if time is running out towards the end of the exam. It is easier to guess and have a correct response for multiple-choice questions than it is for short-answer questions. You'll learn the approach that works best for you when you practise past papers under exam conditions.

Section A: Multiple-choice questions

Choose the best answer

Each question will offer you four alternatives, but only one will be considered correct. Usually one of the options will be obviously wrong, maybe using a term that is nowhere in the course and could even be absurd. Another will be wrong after consideration or analysis. The third possibility may be almost right or may appear correct if you don't read the question carefully. The remaining choice will be the correct answer.

While some choices are very clear, others will involve you having to decide against the incorrect alternatives, known as distractors. In such cases, try to justify why the other three options are not appropriate as well as why the chosen response is correct.

A direct approach to these questions involves reading the question and surveying the answers for the best response on the basis of what you know. If, however, the answer does not stand out, then you should try eliminating the alternatives that are obviously or probably wrong because of specific words or phrases in the suggested answers. This latter approach will improve your chances should you have to give an educated guess for a response.

Beware absolutes and superlatives

Unlike the physical sciences, in which the same reaction will occur every time, psychology is considered an inexact science because it acknowledges all the variables that can affect how behaviour is displayed from one situation to the next. As such, you should avoid responses that use words such as *only*, *every*, *always*, *invariably* and *never*, and choose those with words such as *generally*, *usually*, *ordinarily*, *frequently* and *most often*.

Also be careful if the final alternative is 'all of the above' or 'none of the above'. Don't instantly choose this option. While these responses can sometimes be the correct answer, you must check that *all* the options conform to this answer.

Don't change an answer unless you are sure that it is wrong

Research has shown that your first response is most often the correct one, especially in the case of multiple-choice questions. Only change your answer if you realise that you have misread or missed a key word in the question.

Don't look for patterns in the answers

The answers are randomly spread throughout and will not necessarily include equal amounts of each alternative.

Be careful recording your answers

The answer sheet for this section will be corrected by computer, so you need to be careful when you cross off the desired choice. It is a good idea to mark the chosen alternative clearly in the question booklet and then transfer that choice onto the answer sheet. If you leave a question until later, make sure you also leave it blank on the answer sheet; otherwise it will throw out all of the subsequent answers.

Section B: Short-answer questions

Section B consists of short-answer questions worth different marks.

Questions worth two marks generally focus on the *recall of knowledge* of similar content to notes that you would have in your glossary, such as giving definitions, identifying terms or listing functions or characteristics of a state/condition/phenomenon. When asked to define points or to list their characteristics, you would not be expected to give an explanation that was learned by rote from your textbook. If you understand the concepts, you should be able to use appropriate terminology and expression to give a clear answer in your own words.

Questions worth three marks are designed to assess both knowledge and understanding, and usually involve some application of the theory or analysis of a given scenario. One of the marks awarded in such questions could involve defining or explaining the key concept within the question. The number of marks allocated to the question indicates how many separate points you need to include in your answer.

Extended response questions

The last question in Section B will involve an extended response, usually worth 10 marks. You will be required to respond to the prompt, which will provide a scenario that may relate to any one or more Areas of Study covered within the course, including research methods. Such questions involve applying your theoretical knowledge of particular content covered during the year to the given scenario.

The best approach would be to break down the question into its parts and clearly address each in turn, keeping in mind that there may be a few points to make for some parts of the prompt. Use the marks allocated as a guide to how many pieces of information should be in your answer.

In most cases, the extended response should be in paragraphs, like a mini essay. Doing a brief plan before you begin writing can help you clearly organise your response, while leaving a line between paragraphs helps your assessor to distinguish your points.

Write clearly!

Make sure your handwriting is clear and legible, express yourself coherently and be careful with your spelling. Together, this will help you demonstrate your understanding to the examiners and enhance your potential score for this section.

When asked to explain a term, *avoid* using it as a part of its own definition.

Attempt all questions!

Maximise your potential score. Marks are not deducted for incorrect answers, whereas you might get some points if you make an educated guess. You will definitely not get any marks if you leave a question blank.

Coping with mental blocks

If you come across a question that you cannot answer, don't waste time trying to figure it out. The best thing is to move on to the next question and come back to it later. By doing this, you will be able to do questions for which you *can* get marks, while pondering the earlier question in the back of your mind so that the solution might come to you later.

COMMON COMMAND TERMS USED IN ASSESSMENT

The following words are often used in questions in the end-of-year Psychology exam.

Analyse
To separate and pick out the main points from the information provided

Apply
To use, implement or put the key concepts/knowledge into practice

Cite
To make reference to other material; to refer to the source of material used

Compare
To assess/measure/point out the similarities and differences between two concepts or aspects of key knowledge

Conclude
To reach a deduction based on the information/experimental results

Contrast
To assess/measure/point out the differences between two concepts or aspects of key knowledge

Criticise
To analyse the subject and make judgements, positive as well as negative

Deduce
To arrive at conclusions and/or generalisations based on the given facts

Define
To give a clear, precise, accurate meaning of the term

Demonstrate
To show, explain and describe the relevant concept

Describe
To give a detailed account/informative summary/outline of the key knowledge

Discuss
To argue the pros and cons of a subject or theoretical issue

Distinguish
To name and characterise key pieces of knowledge, so that it is clear how they are different from each other

Evaluate
To judge, assess and weigh the merits of the concept/procedure under scrutiny

Examine
To explore/investigate/review the key knowledge

Explain
To give or clarify the meaning of a concept or the reason for its occurrence

Factors
The facts or circumstances that contribute to a result

Generate
To create and produce statements based on the given information

Graph
To chart/plot given data

Identify
To recognise, detect and point out key concepts in a given scenario or within the information provided

Illustrate
To use a diagram and/or examples to help clarify points under discussion

Implications
Why something is significant or important; long-term effects

Include
To incorporate appropriate material

Interpret
To explain and clarify the key knowledge/data/experimental results

Justify
To show adequate grounds or reasons for conclusions reached

Label
To correctly identify and name tables, graphs, charts and/or diagrams

Limitations
Explanation of how something is not useful or not relevant

List
To give a series of items, steps or concepts

Organise
To arrange or order data or information into a convenient and appropriate form

Outline
To give the main points/facts, leaving out minor details

Recognise
To identify the key knowledge/trends and patterns within the information provided

Relate
To show the relationship between various facts/concepts to connect the present findings to previous research

Research
Systematic investigation of a hypothesis pertaining to an aspect of the key knowledge

State
To present in a clear, concise manner

Suggest
To propose alternative options or directions for further research

Summarise
To abbreviate/condense material in order to give a general account of the main features of the relevant information

Synthesise
To integrate theoretical information and data

Understand
To demonstrate a clear comprehension of the key knowledge, concepts and methods used in research

Use
To apply the key knowledge in a given situation

With reference to
Include this concept or information in your answer

HOW TO USE THIS BOOK

The *A+ VCE Psychology* resources are designed to be used year-round to prepare you for your VCE Psychology exam. *A+ VCE Psychology Units 3 & 4 Study Notes* includes topic summaries of all key knowledge in the VCE Psychology syllabus that you will be assessed on during your exam. Each chapter of this book addresses one area of study. This section gives you a brief overview of the features included in this resource.

Area of Study summaries

The Area of Study summaries at the beginning of each chapter give you a high-level overview of the essential knowledge and key science skills you will need to demonstrate during your exam.

Concept maps

The concept maps at the beginning of each chapter provide a visual summary of each Area of Study outcome.

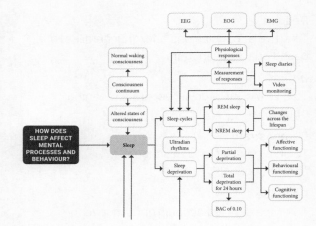

Exam practice

Exam practice questions are at the end of each chapter to test you on what you have just reviewed in the chapter. These are written in the same style as the questions in the actual VCE Psychology exam. There are some official past exam questions in each chapter.

Multiple-choice questions

Each chapter has 17–44 multiple-choice questions.

Short-answer questions

There are 10–24 short-answer questions in each chapter, often broken into parts. These questions require you to apply your knowledge across multiple concepts. Mark allocations have been provided for each question.

Answers

Answers to practice questions are supplied at the back of the book. They have been written to reflect a high-scoring response and include explanations of what makes an effective answer.

Explanations

The answers section includes explanations of each multiple-choice option, both correct and incorrect, and explanations to written response items, which explain what a high-scoring response looks like and signposts potential mistakes.

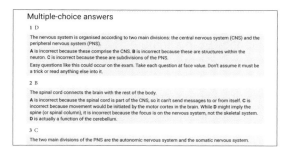

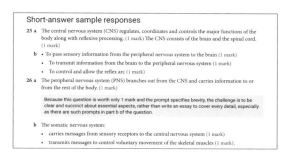

Icons

The icons below occur in the summaries and exam practice sections of each chapter.

Note
Sleep will be explored in more detail in Unit 4 (Chapter 3).

Note boxes appear throughout the summaries to provide additional tips and support.

 ©VCAA 2020 SA Q4

This icon appears with official past VCAA exam questions.

These icons indicate whether the question is easy, medium or hard.

Aboriginal peoples and Torres Strait Islander peoples

Australia is home to two distinct, broad groups of First Nations peoples. These are Aboriginal peoples and Torres Strait Islander peoples. Collectively, the phrase 'Aboriginal and Torres Strait Islander peoples' refers to the group of nations, cultures and languages that live across Australia and throughout the Torres Strait. We speak of 'peoples' (plural) to recognise that there are many different nations, cultures and language groups, not just one Aboriginal and Torres Strait Islander culture or identity.

The term 'Aboriginal' refers broadly to the nations and custodians of mainland Australia and most of the islands, including Tasmania, Fraser Island, Palm Island, Mornington Island, Groote Eylandt, Bathurst Island and Melville Island.

The term 'Torres Strait Islander' refers broadly to the peoples of at least 274 small islands between the northern tip of Cape York in Queensland and the south-west coast of Papua New Guinea.

Visit the Australian Institute for Aboriginal and Torres Strait Islander Studies (AIATSIS) website to view the map of Indigenous Australia, which shows the major language groups and their rough geographical boundaries.

A+ VCE Psychology Practice Exams

 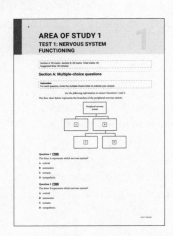

A+ VCE Psychology Study Notes can be used independently or alongside the accompanying resource *A+ VCE Psychology Practice Exams*. *A+ VCE Psychology Practice Exams* features 10 topic tests comprising original VCE-style questions, official VCE questions and two pull-out practice exams. Each topic test includes multiple-choice and short-answer questions, and focuses on one inquiry question of the VCE Psychology syllabus. There are two complete practice exams following the tests. Like the *A+ VCE Psychology Study Notes*, detailed answers are included at the end of the book, demonstrating and explaining how to craft high-scoring exam responses.

A+ DIGITAL FLASHCARDS

Revise key terms and concepts online with the A+ Flashcards.
Just scan the QR code or type the URL into your browser to access them. Note: You will need to create a free NelsonNet account

https://get.ga/
aplus-vce-psych-u34

ABOUT THE AUTHOR

Peter Milesi

Peter Milesi, BBSc (La Trobe), BSW (La Trobe), GradDipEd (Sec.) (ACU), GradDipEdPsych (Monash) Peter has been counselling since 1982, either as a student welfare coordinator or in private practice as a registered psychologist. He has taught Psychology at Years 11 and 12 since its inclusion in the VCE in 1991. Peter has written trial examinations, student workbooks and teacher manuals, and has run revision programs preparing students for semester exams in VCE Psychology.

UNIT 3
HOW DOES EXPERIENCE AFFECT BEHAVIOUR AND MENTAL PROCESSES?

Chapter 1
Area of Study 1: How does the nervous system enable psychological functioning? — 2

Chapter 2
Area of Study 2: How do people learn and remember? — 48

Chapter 1
Area of Study 1: How does the nervous system enable psychological functioning?

Area of Study summary

Unit 3 applies a psychobiological approach to investigate the processes involved in our interactions with our environment.

The first area of study focuses on understanding the responsibilities of specific components within the nervous system that enable us to respond effectively to different stimuli, including how certain neurotransmitters are involved in such responses. This knowledge is then applied within different models to examine the influence stress has on our physical and mental wellbeing.

Area of Study 1 Outcome 1

On completing this outcome you should be able to:

- analyse how the functioning of the human nervous system enables a person to interact with the external world
- evaluate the different ways in which stress can affect psychobiological functioning.

The key science skills demonstrated in this outcome are:

- analyse, evaluate and communicate scientific ideas
- construct evidence-based arguments and draw conclusions.

Adapted from *VCE Psychology Study Design (2023–2027)* © copyright 2022, Victorian Curriculum and Assessment Authority

UNIT 3 / CHAPTER 1 Area of Study 1: How does the nervous system enable psychological functioning?

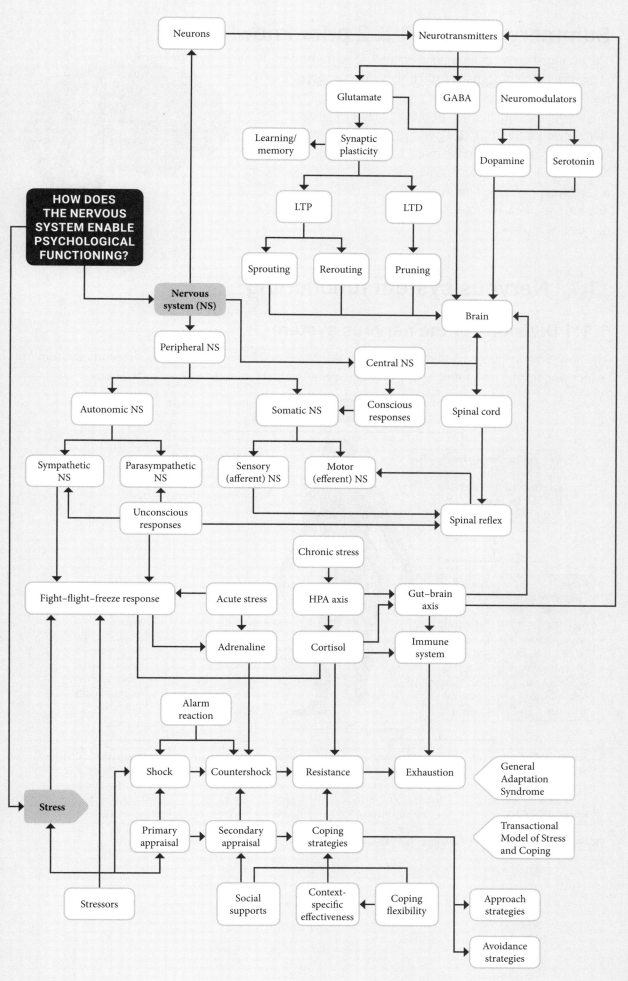

Summary of Units 1 & 2 prior knowledge

The average human brain weighs about 1.4 kilograms and is made up of approximately 100 billion interconnected **neurons** (or nerve cells). These neurons do not function as single entities. Rather, when they are chained or linked into vast networks, they form the nervous system that regulates, coordinates and controls the major functions of the body. Neurons specialise in carrying and processing information and relaying messages between the central nervous system and the peripheral nervous system, as well as activating glands and muscles.

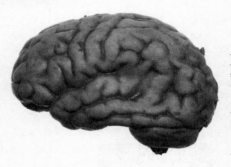

1.1 Nervous system functioning

1.1.1 Divisions of the nervous system

The flow chart in Figure 1.1 summarises the roles of the different divisions of the **nervous system** in responding to, and integrating and coordinating with, sensory stimuli received by the body.

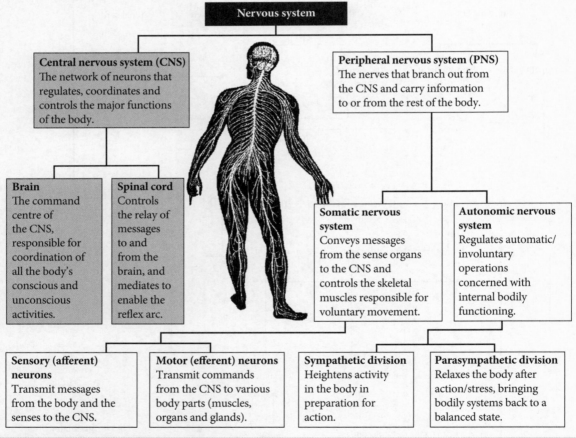

FIGURE 1.1 The divisions of the human nervous system

Central nervous system

The **central nervous system (CNS)** includes the **brain** and the **spinal cord**. The spinal cord plays two major roles in connecting the brain to other areas of the body.

- It transmits sensory information along **sensory neurons** from **receptors** in the body (or the peripheral nervous system) to the brain.
- It transmits information from the brain to the peripheral nervous system, which activates **motor neurons** to enable a response from the skeletal muscles, glands or organs.

Further to this, the spinal cord mediates in the reflex arc to enable an automatic response to certain sensory stimuli.

Peripheral nervous system

The **peripheral nervous system (PNS)** links the central nervous system to other parts of the body and is divided into the somatic nervous system and the autonomic nervous system.

Role of somatic nervous system in control of skeletal muscles

The **somatic nervous system** transports messages between the sense organs and the spinal cord. Via its connections with the skeletal muscles, it also controls voluntary bodily movements. Nerves from the somatic nervous system do *not* control the non-skeletal muscles, such as the heart, lungs, stomach or intestines.

Within the somatic nervous system, sensory and motor neurons perform quite distinct functions.

Sensory (afferent) nerve fibres receive information from sensory stimuli arising inside or outside the body and transmit messages inwards to the central nervous system, where they are correlated. The motor (efferent) nerve fibres transmit commands outwards, away from the central nervous system, enabling the coordination of organs such as muscles and glands. This allows them to work harmoniously, thus ensuring the wellbeing of the individual.

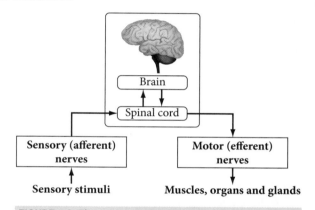

FIGURE 1.2 The somatic nervous system

Note
To remember the different names for the two types of nerves, use the acronym **SAME**, which stands for Sensory (Afferent) neurons and Motor (Efferent) nerves.

Conscious and unconscious responses

Activity of the somatic nervous system is usually associated with **conscious responses**; that is, behaviours involving attention and awareness, such as actively reading this book and then picking up a pen to write down some notes. An individual actively processes sensory input within the brain to make intentional decisions about an appropriate course of action before then sending commands via efferent neurons to enable a controlled reaction to stimuli.

Unconscious responses, on the other hand, do not involve our immediate, direct awareness in order to occur. Such responses are usually automatic reflexes that 'bypass' the brain to speed up reactions to be more efficient and/or increase our survivability.

Spinal reflex

One of the functions of the spinal cord is to mediate between the divisions of the somatic nervous system to enable the **spinal reflex** (or reflex arc), which involves an automatic response to certain sensory stimuli. Such a reflex occurs without any assistance from the brain, although the brain is informed, usually registering it after this action has taken place. For example, if you touched the side of a hot kettle, you would have already pulled back your hand before you actually register the pain.

Pain (sensory) receptors within the skin are stimulated, triggering an impulse (action potential) to pass along the sensory neuron to the spinal cord. The impulse then splits, passing through the **interneuron** within the spinal cord that connects to a motor neuron as well as sending an impulse through the ascending **afferent pathways** to the brain. The impulse from the interneuron proceeds along the motor neuron to effector cells, which respond by contracting the skeletal muscles to withdraw from the painful stimulus.

The brain receives and processes the impulse from the ascending afferent pathways in order to consciously register the sensation and that the reflex has taken place.

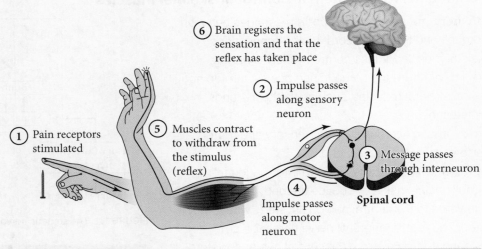

FIGURE 1.3 The spinal reflex

Role of the autonomic nervous system in control of non-skeletal muscles

The **autonomic nervous system (ANS)** controls *involuntary* bodily activities, such as digestion and heart rate, where nerves transmit information to and from the glands and internal organs in order to enable a balanced state.

When the fight-or-flight response is activated, the **sympathetic nervous system** is turned on, initiating some responses and inhibiting others, creating a state of arousal to mobilise the body to deal with an emergency situation.

Once the emergency is over, the **parasympathetic nervous system** conserves bodily functions by calming the body, restoring it to a balanced state of homeostatic equilibrium.

> **Note**
> To remember the different divisions of the autonomic nervous system, associate the **s**ympathetic branch with **s**tress and/or arou**s**al.
>
> Imagining a **para**chute floating down to the ground may help you to remember the prefix for the **para**sympathetic branch, which slowly brings physical mechanisms back down to normal levels again, relaxing the body after arousal.

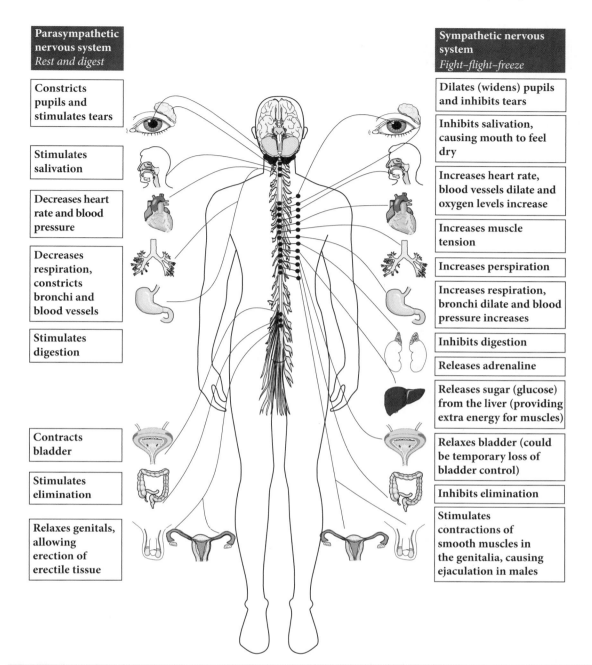

FIGURE 1.4 The main functions of the sympathetic and parasympathetic nervous systems

Note
Other mnemonics for each division are the rhymes **'fight and flight'**, referring to the physiological arousal triggered by the sympathetic nervous system, and **'rest and digest'** or **'feed and breed'**, which acknowledge the functions initiated by the parasympathetic nervous system once the threat has passed.

1.1.2 Role of neurotransmitters in transmission of neural information across a neural synapse

All neurons transmit information in the form of nerve impulses involving electrochemical reactions known as the **action potential**, which occur due to the flux of ions across the membranes as they travel along the axons. The electrical activity within the axon is insulated by a fatty layer known as the myelin sheath. The transmission of information from one neuron to the next, however, requires the release of specialised **neurotransmitters**, which are a variety of different chemicals capable of altering the electrical activity, or interior charge, in other neurons.

Neurotransmitters are manufactured and stored inside synaptic vesicles. The arrival of neural impulses at the axon terminal (also referred to as the synaptic knob) triggers these vesicles to move to the surface of the presynaptic neuron and release the appropriate neurotransmitters.

These neurotransmitters cross the synaptic gap (or cleft) between neurons. This is a microscopic space between the presynaptic axon terminal and the postsynaptic dendrite of the next neuron that changes its activity after the neurotransmitter attaches to specific receptor sites, which are areas on the surface of the receiving neuron that are sensitive to both neurotransmitters and hormones. The unique structure of each neurotransmitter molecule means that it fits into specifically shaped complementary receptor sites in the way that a key fits perfectly into a lock. Depending on the chemical reaction at these dendritic receptor sites, there could be an excitatory effect, which sets up an action potential that is passed as an electrical impulse along the axon to the next set of **synapses**, or an inhibitory effect, which prevents the action potential.

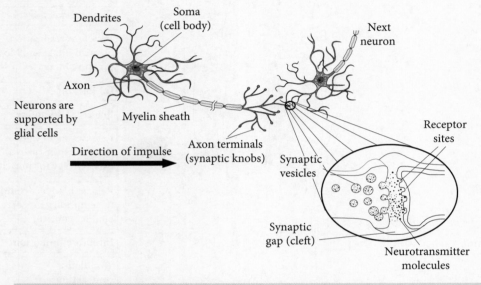

FIGURE 1.5 The transmission of neural information

Agonists are other chemicals that have a similar molecular shape, allowing them to fit into the receptor sites. As they are chemically similar, they activate the receptors, acting like the neurotransmitter itself. Agonists may be employed to supplement specific neurotransmitters if they are lacking.

Antagonists, on the other hand, are chemicals that can fit into the receptor site but will not activate them, thereby blocking the neurotransmitter and inhibiting its function. Antagonists limit the effect of specific neurotransmitters if there is an excess.

Excitatory and inhibitory effects of neurotransmitters

Postsynaptic neurons have many different receptor sites capable of receiving several distinct neurotransmitters. The effects of a neurotransmitter are not entirely caused by the chemical itself, but could be modified by the combined interaction it may have with other neurotransmitters that are also being taken in by other receptors on the postsynaptic neuron.

Excitatory neurotransmitters

Excitatory neurotransmitters are the nervous system's 'on switches', speeding up neural activity and increasing the likelihood that an excitatory signal is sent. Excitatory transmitters regulate many of the body's most basic functions, including thought processes, the fight-or-flight response, motor movement and higher thinking. Physiologically, the excitatory transmitters act as the body's natural stimulants, generally serving to promote alertness, energy and activity. However, without a functioning inhibitory system, to put on the brakes, things can get out of control.

Excitatory neurotransmitters include glutamate, dopamine, norepinephrine and acetylcholine.

Glutamate is a major excitatory neurotransmitter and is used at the great majority of fast excitatory synapses in the brain and spinal cord. It is also used at most synapses that are 'modifiable', and is thereby associated with **long-term potentiation**, creating links between neurons that form the basis of learning and long-term memory. An excess of glutamate is associated with **anxiety** disorders, attention deficit hyperactivity disorder (ADHD) and even seizures.

Excitation in the brain must be balanced with inhibition. Too much excitation can lead to restlessness, irritability, insomnia and even seizures.

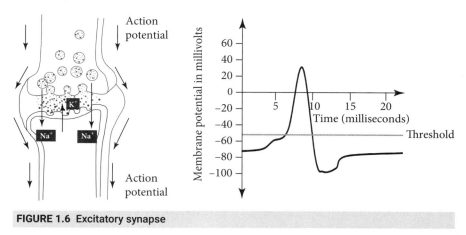

FIGURE 1.6 Excitatory synapse

Inhibitory neurotransmitters

Inhibitory neurotransmitters are the nervous system's 'off switches', decreasing the likelihood that an excitatory signal is sent. These chemical messengers slow down neural activity by making a postsynaptic neuron less likely to fire an action potential. Physiologically, the inhibitory transmitters act as the body's natural tranquilisers, generally serving to induce sleep, promote calmness and decrease aggression.

Inhibitory neurotransmitters include GABA and serotonin.

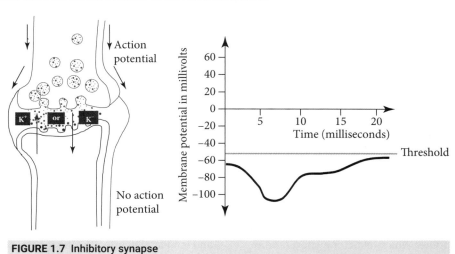

FIGURE 1.7 Inhibitory synapse

GABA is the abbreviation for **gamma-aminobutyric acid**. GABA is the major inhibitory neurotransmitter in the central nervous system and plays a major role in regulating anxiety and reducing stress by having a calming effect on the brain. It improves mental focus while calming the nerves.

GABA acts like a brake to the excitatory neurotransmitters (norepinephrine, adrenaline, dopamine and serotonin), which can cause anxiety if the system becomes overstimulated.

1.1.3 Role of neuromodulators and their effect on brain activity

Many chemical messengers that are neurotransmitters also act as **neuromodulators**, which can have a range of effects on brain activity.

Unlike neurotransmitters, these molecules are not reabsorbed by the presynaptic neuron or broken down by enzymes. This means that they spend a significant amount of time in the cerebrospinal fluid, influencing (or 'modulating') the activity of neurons in the brain, acting together with neurotransmitters to enhance the excitatory or inhibitory responses of receptors.

Neuromodulators initiate a second messenger within the postsynaptic cell, which generates a longer, more sustained signal that can lead to lasting changes in cellular activity associated with long-term potentiation to improve synaptic transfer.

As the release of neuromodulators occurs in a diffuse manner ('volume transmission'), they are not restricted to a specific synaptic cleft between two neurons, and so can affect large numbers of neurons at once, altering the activity of a brain region rather than just an individual neuron. This general release into the neural tissue also means that neuromodulators have a very long range of action compared to regular neurotransmitters, as they can affect neurons quite far from the site of release, not just influence cells near the presynaptic neuron.

TABLE 1.1 Similarities between neurotransmitters and neuromodulators

- Both are types of chemical messengers released by the nervous system.
- They are stored within synaptic vesicles in axon terminals of the presynaptic neuron and released into the synaptic cleft.
- They transmit neural impulses through the synapse, binding to specific receptors on the postsynaptic neuron or effector cells.
- Their effect can be either excitatory or inhibitory.

TABLE 1.2 Differences between neurotransmitters and neuromodulators

Neurotransmitters	Neuromodulators
Chemical molecules released from the axon terminal of a presynaptic neuron to diffuse across the synaptic cleft to send signals to another postsynaptic neuron or an effector cell.	Chemical molecules released from the axon terminal of the presynaptic neurons, which alter the strength of neuronal signal transmission by controlling the number of neurotransmitters synthesised and released in response to stimuli.
Action occurs at a specific synapse during direct synaptic transmission, thereby affecting one or two postsynaptic targets (neurons or effector cells).	Action occurs through volume transmission, affecting a diverse group of postsynaptic targets (neurons or effector cells) with the appropriate receptor.
Transmission of chemical signals to adjacent postsynaptic targets.	The site of action can be either near the site of release or quite far away from it.
Act directly on their postsynaptic target by binding to specialised receptors to cause a specific response.	Indirectly affect their postsynaptic targets via second messengers.
Degraded or rapidly recycled by the presynaptic neuron (reuptake).	Not degraded rapidly or reabsorbed by the presynaptic neuron.
Produce a rapid effect to pass the signal from one neuron to another, which lasts for a short time.	Prolonged activation of target cells produces a slow and long-lasting effect to change the cellular or synaptic properties of neurons.

Dopamine

As there are several **dopamine** pathways in the brain, this chemical messenger is involved in many diverse functions, depending on the region of the brain involved. Within the midbrain and basal ganglia, dopamine is necessary for the fine-tuning of motor control to allow smooth, coordinated movement. It allows us to be alert and attentive, enabling cognitive processing within our working (short-term) memory.

Dopamine is part of the brain's reward system and creates feelings of satisfaction or pleasure when we do things we enjoy. It is therefore involved in motivating us to behave in certain ways to achieve these rewards and to feel good when we receive them. It is this system that is prevalent in all forms of compulsion and addiction.

Variations in this neuromodulator function appear to be associated with variations in personality, resulting in changes in relatively stable patterns of behaviour, motivation, emotion and cognition. In excess, dopamine has been implicated in anxiety, aggression, addiction and schizophrenia. Deficiencies in dopamine have been linked to a lack of motivation and desire, motor problems and rigidity (such as in Parkinson's disease), and difficulties with focusing attention and cognitive impairment ('brain fog').

Serotonin

Serotonin is used in various parts of the nervous system. While it is most well-known for its role in the brain, where it plays a major part in mood, anxiety and happiness, serotonin is also involved in sleep, memory processing and appetite.

Serotonin's influence on regulating mood makes it integral to our overall sense of wellbeing. Reduced quantities of serotonin in the brain are associated with anxiety and several mood disorders, including depression. Selective serotonin reuptake inhibitors (SSRIs) are widely used antidepressants as they temporarily prevent the removal of serotonin from specific synapses, thereby enhancing the effect of released serotonin to elevate mood.

Serotonin plays a key role in the enteric nervous system and the gut–brain axis. The majority of the body's serotonin (80–90%) is produced in the gastrointestinal tract, where it has a role in promoting healthy digestion and regulating bowel function and movements. It also plays a part in reducing the appetite while consuming a meal.

Along with other neurotransmitters such as dopamine, serotonin can influence the pattern and quality of our sleep, as it is required to make melatonin, a hormone critical to the proper functioning of our sleep–wake cycle.

> **Note**
> Sleep will be explored in more detail in Unit 4 (Chapter 3).

Table 1.3 summarises the role of some of the neurotransmitters mentioned in the VCE Psychology course. Those that can also act as neuromodulators have been indicated with an asterisk (*).

TABLE 1.3 Key neurotransmitters and their functions

Neurotransmitter	Major functions
Glutamate (Glu)	Main agonist (excitatory) neurotransmitter in the brain, speeding up neural activity; stimulates hypothalamus; associated with long-term potentiation (learning and long-term memory).
GABA (gamma-aminobutyric acid)	Main inhibitory neurotransmitter in the brain, slowing down neural activity.
Dopamine (DA)*	Voluntary movement; learning and memory; stimulates hypothalamus; alertness and attention; emotion; drive/motivation; pleasure/reward-seeking.
Serotonin (5-HT)*	Sleep–wakefulness cycle; mood regulation; hunger and appetite; impulsivity; sensory perception; pain sensitivity and suppression; temperature regulation.
Noradrenaline (norepinephrine)*	Stress-related activation of sympathetic nervous system: fight-or-flight response; awakening, alertness and arousal; attention and concentration; emotion; mood elevation; learning and short-term memory.
Acetylcholine (Ach) *	Attention and arousal; promotes rapid eye movement (REM) sleep; cognitive functioning; learning and memory – prevalent in the hippocampus; autonomic nervous system; works to activate muscles for voluntary movement.

1.1.4 Synaptic plasticity

As the brain receives specific appropriate input through the senses or processes information, with appropriate frequency, intensity and duration, it changes its physical structure. This structural change takes place through the formation and alteration of neural pathways or networks between the brain's neurons, which are operative in learning and memory.

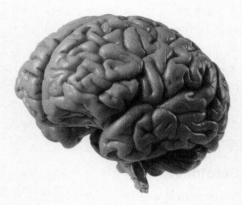

Neuroplasticity refers to the brain's ability to change and reorganise itself by forming new neural connections throughout life. According to the theory of neuroplasticity, thinking, learning and behaviour actually change both the physical structure of the brain's anatomy and its functional organisation.

Changes associated with learning occur mostly at the synaptic level, where new connections can form and the internal structure of the existing synapses can be modified to make synaptic transmission more efficient. Neural pathways are reorganised based on experience to incorporate new behaviours that are more able to respond to the demands of our environment. Such adaptations are more likely to help preserve a living organism and improve its chances of survival.

Plasticity within the brain occurs:
- at the beginning of life, when the immature brain organises itself
- in cases of brain injury to compensate for lost functions or to maximise remaining functions
- throughout adulthood, whenever something new is learned and memorised.

Neuroplasticity forms the scientific basis for treatment of acquired brain injury, with goal-directed rehabilitation programs aiming to manage the functional consequences of the injury.

TABLE 1.4 Types of neuroplasticity

	Developmental plasticity	**Adaptive plasticity**
Definition	The brain's natural ability, especially during infancy, to change neural structure and form new connections in order to process sensory information from environmental stimulation and experiences during its growth and development.	The brain's ability to reorganise and change neural structure, forming new connections in order to relocate or reassign functional areas within the brain. Such changes enable adjustment to experiences, to recover or compensate for lost function and/or to maximise remaining functions in the event of brain damage or injury, such as traumatic head injury or stroke.
Neuronal changes	Proliferation Neural migration **Synaptogenesis** Synaptic pruning Myelination	Rerouting Sprouting Pruning
Determined by	Genetically predetermined; occurs in response to the initial processing of sensory information by the immature brain.	May be environmentally determined; for example, compensating for acquired brain injury or in adjustment to new experiences.
When does it occur?	Over the life span, but diminishes with age.	Also occurs over the life span, but it is more efficient and effective during infancy/early childhood.

The concept of neuroplasticity associated with experience-driven alterations in synaptic structure and function was central to the research by Eric Kandel. This research verified that memory circuits within the neurons of the *Aplysia*, a genus of sea slug, were created and modified as a result of functional and structural changes at the neuronal level when synapses strengthen or weaken over time in response to increases or decreases in their activity.

Kandel and his colleagues were able to identify, through stimulation, various types and groups of neurons involved in the *Aplysia*'s behaviour, allowing them to examine which cellular components of the neural circuits changed during learning, as well as the molecular mechanisms responsible for the changes in neuronal function.

In repeatedly activating the *Aplysia*'s protective siphon-withdrawal reflex, the cellular structure of the neurons was altered, increasing the number of interconnecting dendrites and axon terminals to form more communication points (synapses).

Continued stimulation strengthened the synaptic connections in the same circuit of neurons and their function improved by becoming more efficient, resulting in an increased release of neurotransmitters. As a result, the slug was able to avoid the negative stimuli that had initially caused the reflex.

Kandel showed that weak stimuli give rise to certain chemical changes in synapses which form the basis of short-term memory. In contrast, stronger stimuli, or repeated exposure to the stimulus in question, cause an increase in the number of synapses, along with different synaptic changes requiring protein synthesis that produce an alteration in the shape of synapses found on neuronal membranes. Kandel found that these new synaptic shapes increased the sensitivity of synapses to further stimulation and improved synaptic function, which resulted in demonstrable long-term memory.

Neural pathways involved in learning

In order to learn or memorise a fact or skill, there must be persistent functional changes in the brain that represent the new knowledge. For example, when you are exposed to a new concept, you have to make new connections between particular neurons in your brain to process it. To learn this new concept, you would reinforce your understanding several times. By selecting and strengthening the connections within the various circuits in your cortex, you enable a new, durable association among these neurons to form your memory of that concept. To remember this information at a later date, you will have to successfully reactivate these same neural circuits. This will be easier if, when you first learned the material, these circuits were strengthened by revising the concept and thus sending the corresponding nerve impulses down them many times. In contrast, if you repeated the concept only a few times, then the neural connections would be weaker, and the circuit would be harder to reactivate.

The brain can thus be said to store information in networks of modified connections between neurons, the arrangement of which constitutes the information, and the activation of which constitutes retrieval of this information. Multiple memories can be encoded within a single neural network by different patterns of synaptic connections. Conversely, a single memory may involve simultaneously activating several different groups of neurons in different parts of the brain.

Synaptic changes

Learning, therefore, is a function of transmission of signals along the neurons and across their interconnections. At the synapse, nerve impulses are converted to chemical processes that excite activity in the connecting neuron, thereby strengthening and possibly creating new synaptic connections, or inhibit activity in the connecting neuron, weakening the connections, leading to the deterioration of existing synapses.

In attempting to explain learning and memory as a function of cellular change, Donald Hebb proposed that if two neurons are active at the same time, the synapses between them are strengthened, thus facilitating the passage of nerve impulses along a particular circuit. He maintained that the repeated firing of axons results in metabolic changes in the presynaptic and postsynaptic neurons, reinforcing the connection between them. In other words, learning produces lasting chemical changes in nerve cells, which make it easier for connected neurons to communicate with each other, and therefore form memories. Hebb's hypothesis led to the discovery of long-term potentiation (LTP) in the early 1970s. More recent research has uncovered that the LTP that leads to synaptic strengthening is very specific to the network of neurons that are activated simultaneously, and only to such neurons.

Successful learning is a function of establishing new synaptic connections to bring about changes in the structure of existing neural networks. The formation of new synapses depends on stimulation through activity.

> **Note**
> Neurons that fire together, wire together!

TABLE 1.5 Comparison of long-term potentiation and long-term depression

	Long-term potentiation (LTP)	**Long-term depression (LTD)**
Description	The long-lasting strengthening of synaptic connections of neurons, resulting in enhanced or more effective functioning of the neurons whenever they are activated, thereby improving the ability of two neurons to communicate with one another at the synapse.	In opposition to LTP, LTD serves to selectively weaken specific synapses in order to make constructive use of synaptic strengthening caused by LTP. This is necessary because, if allowed to continue increasing in strength, synapses would ultimately reach a limit to their efficiency, which would inhibit the encoding of new information. Hippocampal LTD may be important for the clearing of old memory traces, which contributes to the decay of memory.
Associated neuronal changes	Synaptogenesis Circuit formation	Synaptic pruning Circuit pruning
Determined by	More neurotransmitters are produced and released by presynaptic neurons, which act on the receptor sites of postsynaptic neurons. Because of repeated stimulation, the postsynaptic neurons become more and more responsive to the presynaptic neurons.	Generally, the mechanism of LTD occurs due to reduced activation of the neural pathway as a result of low-frequency stimuli or a lack of stimulation and a slow rise in postsynaptic calcium.

These processes work together to modify connections between neurons as the fundamental mechanism of memory formation that enables learning to take place.

> **Note**
> These processes can be summed up by the phrase 'use it or lose it'. Synapses that are regularly activated will be strengthened and retained, whereas those that are not will be eliminated.

A number of studies across several different species have independently shown consistent results demonstrating the neural changes associated with memory. Following stimulation and repeated activation, the structure of the neurons was altered by growing more branches (dendrites and axon terminals) to make more connections (synapses) with other neurons. This process is referred to as **sprouting**.

In cases where damage has occurred within the brain, undamaged neurons may sprout new extensions to compensate for any loss of function. In addition, if the undamaged neuron has lost a communication point with an active neuron, it may seek a new active neuron and link with it instead. In seeking out and connecting with other neurons, **rerouting** occurs to allow messages to be sent along a new neural pathway to go around the damaged area.

In addition to this, if a synapse is not regularly activated, the connection will weaken and ultimately be removed. This process is known as **synaptic pruning**. By removing unnecessary connections, the neuron becomes more efficient and able to form further, more productive networks with other neurons.

> **Note**
> Gardeners prune plants by cutting away dead or useless branches to open up the plant and promote healthier growth. Synaptic pruning is similar in that redundant connections are removed to allow the neuron to grow and make more useful links with other neurons.

Glutamate's role in synaptic plasticity

As the main excitatory neurotransmitter throughout the brain and central nervous system, glutamate enhances information transmission by initiating activity in postsynaptic neurons, making them more likely to fire an action potential.

Synaptic plasticity is usually dependent on the effect of glutamate, which can change the efficacy of a particular synapse by increasing or decreasing the amount of neurotransmitter released across the synapse or by increasing or decreasing the amount of receptors present. Furthermore, the changes that occur at one synapse affect the entire network of neurons to which that synapse is connected.

Glutamate is responsible for the different forms of synaptic plasticity such as **long-term potentiation (LTP)** and **long-term depression (LTD)**, mechanisms that are thought to underlie learning and memory. Repeated stimulation of the postsynaptic neuron will increase the number of glutamate receptors on the cell, leading to LTP, and can initiate growth of new dendritic spines on postsynaptic neurons, resulting in the formation of additional synaptic connections.

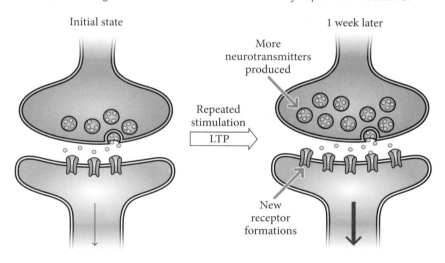

FIGURE 1.8 The effect of neurotransmitters in long-term potentiation

The opposite is also true. Low frequency or a lack of stimulation would decrease the number of glutamate receptors leading to LTD of that cell, which could prune the number of dendritic spines, resulting in reduced synaptic connections.

In memory formation, glutamate plays crucial roles in the structural changes that occur, particularly in the growth and strengthening of synaptic connections between neurons within a memory circuit. Specific types of glutamate receptors have been found to be abundant in the **hippocampus**, a key structure involved in the formation of long-term memory.

1.2 Stress as an example of a psychobiological process

Stressors are circumstances that are perceived as threatening or challenging to a person's wellbeing or which exceed their ability or resources to cope, such as daily pressures, life events and trauma. **Stress** involves prolonged levels of physiological and psychological arousal associated with the fight-or-flight response, when the sympathetic nervous system is activated to prepare or mobilise the body to deal with one or more stressors.

Stress can have positive effects as well as negative ones. Even though stress can motivate us into action and can enhance our performance in certain tasks (**eustress**), most people focus on the negative and harmful effects of stress (**distress**).

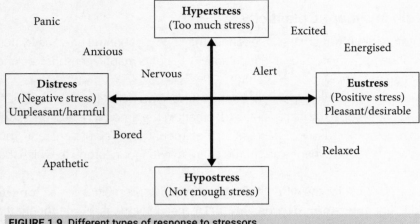

FIGURE 1.9 Different types of response to stressors

Acute stressors are challenging situations that trigger an immediate response and are usually short-lived, as they are dealt with and resolved quickly. **Chronic stressors**, on the other hand, involve situations that are ongoing and lead to prolonged levels of physiological arousal.

1.2.1 Stress response

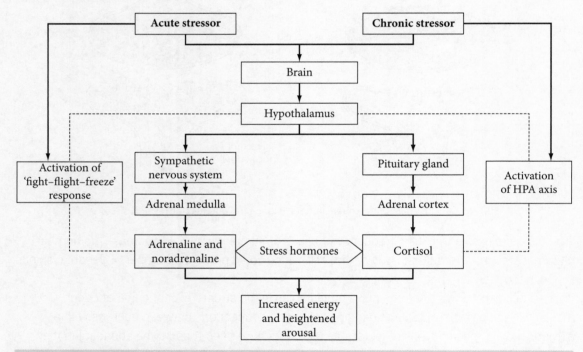

FIGURE 1.10 The physiological systems involved in stress

The role of **adrenaline** and **noradrenaline** is to stimulate the sympathetic nervous system to activate the fight-or-flight response and raise physiological levels of arousal in order to deal with an immediate **challenge**.

Fight–flight–freeze response

While initial theories emphasised the fight-or-flight response to perceived danger, current models recognise that survivability may also be enhanced by the 'freeze' response. All three responses in the **fight–flight–freeze response** are considered adaptive survival mechanisms that enable us to react effectively to a **threat** and help minimise harm to our wellbeing.

Fight-or-flight response

The physiological responses associated with stress are produced when the fight-or-flight response is triggered by the hypothalamus, activating the sympathetic division of the autonomic nervous system. The fight-or-flight response refers to a state of arousal that prepares or mobilises the body to confront a situation or 'fight' it, or run away (flee) from a situation, hence the term 'flight'.

Typical responses include:

- increased levels of adrenaline (thus speeding up heart rate and response time)
- increased heart rate and blood pressure (to quickly carry more oxygen and glucose to the body)
- increased respiration rate (to supply more oxygen to the blood)
- increased muscle tension (to enable quicker reaction time)
- release of sugar from the liver (to supply energy to the brain and skeletal muscles)
- dilated (widened) pupils (to allow more information to be gathered quickly)
- dry mouth (due to saliva inhibition)
- inhibited digestion (oxygen is diverted to muscles for a faster response) causing 'butterflies' in the stomach
- sweating (leading to increased electrical conductivity of the skin)
- forming of 'goosebumps' on the skin
- an increased need to urinate (due to a relaxed bladder).

'Freeze' response

If the threat is perceived to be too intense or too much to deal with, the individual (or animal) may become overwhelmed, feeling like there is no way to respond, no chance for survival or no chance for escape. In such cases, the 'freeze' response may be initiated to reduce the exposure to the perceived threat.

In this response, the parasympathetic nervous system takes over to overshadow the existing effects of the sympathetic nervous system activation. Physiological effects include significant changes, including a slowing heart rate, a drop in blood pressure and a loss of muscle tone that prevents movement. Despite this, the body is still in a state of high arousal as *both* divisions of the autonomic nervous system are activated. As such, the individual is conserving energy while also being in a mobilised state ready for action. This allows them to flee from the threat very quickly by shifting back to a full sympathetic response when there is a chance for escape after having been in a frozen state.

Role of cortisol in chronic stress

Because of the increased 'wear and tear' on the body due to prolonged stress, **cortisol** is released to reduce inflammation by blocking the activity of the white blood cells. It also helps repair muscles and speeds up the healing process. In addition, cortisol aids in the metabolism of fat, protein and carbohydrates, and stimulates the formation of glucose to increase blood sugar levels to provide additional energy for muscles.

However, sustained elevated levels of cortisol in the bloodstream as a result of prolonged chronic stress can suppress the **immune system**, reducing its effectiveness to combat harmful bacteria and making the individual vulnerable to stress-related illnesses.

While the release of stress hormones helps to speed up systems in the body to maintain alertness and help the body to fight the stressor, prolonged release may lead to permanent changes in heart rate and blood pressure, increasing an individual's likelihood of contracting heart disease or having a stroke.

1.2.2 Gut–brain axis

The **gut–brain axis (GBA)** refers to the two-way connection and communication pathway between the gut microbiome and the brain. Your brain influences your gut health, and conditions in the gut can directly affect brain function and our mental health.

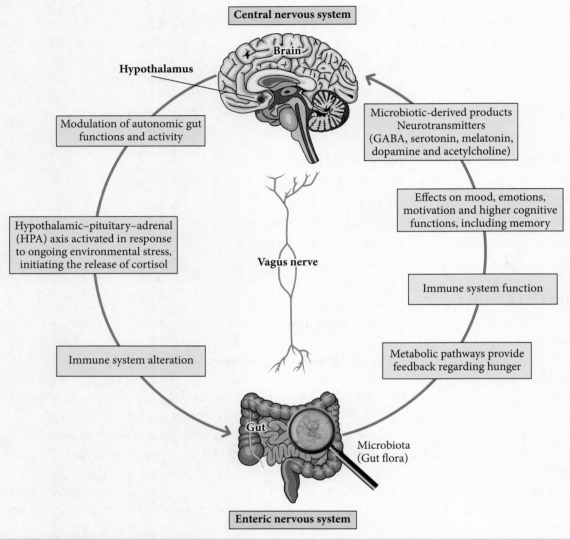

FIGURE 1.11 The gut–brain axis

Emerging research into the microbiome GBA has revealed a complex network that integrates the central nervous system, neuroendocrine and immune systems, including the **hypothalamic–pituitary–adrenal (HPA) axis**, both divisions of the autonomic nervous system, including the enteric (gastrointestinal) nervous system and the vagus nerve, and the gut microbiota.

The **enteric nervous system**, which governs the function of the **gastrointestinal system**, is connected to the brain through the **vagus nerve**, which sends signals in both directions to connect, via direct and indirect pathways, cognitive and emotional centres in the brain with peripheral intestinal functions.

Descending **efferent pathways** from the brain send messages to the gut to monitor and integrate autonomic gut functions, ensuring the proper maintenance of gastrointestinal **homeostasis** and mechanisms such as immune activation.

Ascending afferent pathways carry feedback from the intestines to the **hypothalamus** (which controls basic drives associated with hunger and emotions) and limbic system (which regulates emotions and memory), which can have differing effects on mood, motivation and higher cognitive functions.

The concept of a microbiome GBA has now been expanded to include the potential role of gut microbiota in influencing brain function, behaviour and mental health.

Gut microbiota (or **gut flora**) is an assortment of micro-organisms inhabiting the mammalian digestive tract. The composition of this microbial community is host specific, changing throughout an individual's lifetime and susceptible to both external and internal factors, such as when diet varies or when there is a change in one's overall health.

Research evidence suggests that enteric **microbiota** have an important impact on the nervous system, interacting directly with the central nervous system by influencing brain chemistry and affecting neuroendocrine systems associated with the stress response, anxiety and memory function.

Many neurotransmitters are produced by your gut microbiota, such as acetylcholine, dopamine, GABA, melatonin and serotonin, which are essential for regulating peristalsis and sensation in the gut. A large proportion of serotonin, which regulates mood and cognition and also helps control your body clock, is produced in the gut. Many species of *Lactobacillus* and *Bifidobacterium* in the gut produce GABA, which is the main inhibitory neurotransmitter in the brain that helps control feelings of fear and anxiety.

Psychosocial factors influence the actual physiology of the gut. Messages from the limbic system (activated by various types of emotion or stress) influence autonomic activity of the gut, which can often trigger a number of symptoms, such as feeling 'butterflies' in your stomach.

Chronic environmental stress activates the HPA axis that, in turn, leads to cortisol release from the adrenal glands. Increased cortisol levels cause changes in the gut microbiota and intestinal epithelium, and can have wide-ranging systemic effects, especially on the body's immune system.

Emerging research data support the role of microbiota in influencing anxiety and depressive-like behaviours. Animal studies have demonstrated that microbiota influence stress reactivity and anxiety-like behaviour, and regulate the set point for HPA activity. Further to this, other studies have shown that stress inhibits the signals sent through the vagus nerve and also causes gastrointestinal problems.

Such findings highlight the importance of preserving a proper balance in our gut bacteria in order to maintain adequate mental wellbeing.

1.2.3 General Adaptation Syndrome

One of the first psychologists to recognise the relationship between stress and disease was Hans Selye, who proposed the **General Adaptation Syndrome (GAS)**. He stated that the body seems to respond in the same negative manner to all different kinds of stressors, whether they are positive or negative, and that the body goes through three different stages when confronting ongoing stress.

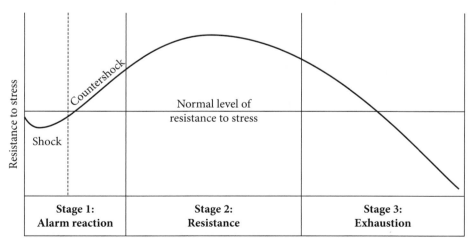

FIGURE 1.12 The General Adaptation Syndrome

In the **alarm reaction stage**, we are faced with a new stressor. Our body temporarily goes into **shock** and our resistance to stress drops below normal levels. Consequently, our blood pressure and body temperature drop, and our muscle tone is affected. **Countershock** then occurs when the sympathetic nervous system activates, mobilising bodily resources to cope with the stressor. The main response at this stage is like that of the fight-or-flight response; that is, the organism becomes highly aroused and alert as it prepares to deal with the stressor. Adrenaline and noradrenaline are released into the bloodstream, speeding up our heart and respiratory system in response. Symptoms such as headaches and stomach aches may be experienced. Cortisol is also released to increase glucose in the bloodstream to provide additional energy for muscles.

During the second stage of the GAS, the **resistance stage**, the body adjusts to the stress through allostasis and the normal level of coping rises. But this is at a high physical cost. The body has now become susceptible to other stressors. Prolonged release of cortisol interferes with the body's immune system. Ability to withstand infection is lowered over time, and the first signs of psychosomatic disorders begin to appear.

In most instances, stressors encountered are quickly and effectively dealt with, enabling the organism's state to eventually return to normal. However, when efforts are unsuccessful and the individual cannot adapt to continuing stress, or if there are additional stressors causing a cumulative effect, then they will enter the **exhaustion stage**. In this final stage, bodily resources are now depleted and resistance to stress has dropped dramatically below normal levels.

While stress does not necessarily cause disease, a strong relationship has been established between chronic stress and the increased likelihood of contracting a psychosomatic illness or making a current illness worse. This is because, over time, the overexposure to stress can have adverse cumulative effects on various organ systems, leading to the inhibition of the immune system and disease.

While cortisol helps to repair the body and speed up the healing process, the immune system attempts to reduce the levels of these hormones – akin to 'an army fighting on two fronts' – thereby weakening it. As a result, harmful bacteria are not recognised and eliminated from the body at a fast-enough pace. Therefore, the activities of lymphocytes (which discover and destroy harmful cells) and phagocytes (which ingest and eliminate these cells) are suppressed, making the body more prone to infectious diseases.

Unless some stress management is sought, **psychosomatic illnesses** could occur. These medical problems may be major or minor in nature and stem from a combination of psychological, emotional and physical factors. Examples of psychosomatic conditions include hypertension, high blood pressure, cardiac problems, fatigue, asthma, arthritis, backache, constipation, eczema, diarrhoea, stomach pain, hives, indigestion, insomnia, migraines, muscle soreness and ulcers. In extreme cases, complete collapse may occur.

> **Note**
> To recall the different stages of the **G**eneral **A**daptation **S**yndrome, remember the sentence: 'The **GAS** blast gave me a **SCARE**.' (**S**hock and **C**ountershock occur in the **A**larm stage, followed by **R**esistance and then finally **E**xhaustion.)

1.2.4 Transactional Model of Stress and Coping

While Selye's General Adaptation Syndrome views stress as a fairly automatic physiological response to any threatening stimulus, Lazarus and Folkman's **Transactional Model of Stress and Coping** emphasises the meaning that an event has for the individual. According to this model, stress is the product of a transaction between a person and their external environment in their evaluation (appraisal) of a stressor.

Stress is highly subjective. Two people may view the same situation very differently. For example, one student may see singing a solo at a school assembly as very exciting, while another may find just the thought of it enough to lead to an anxiety attack.

When faced with a stressor, a person evaluates the potential threat (**primary appraisal**), making a judgement about the significance of an event as stressful, positive, controllable, challenging or irrelevant.

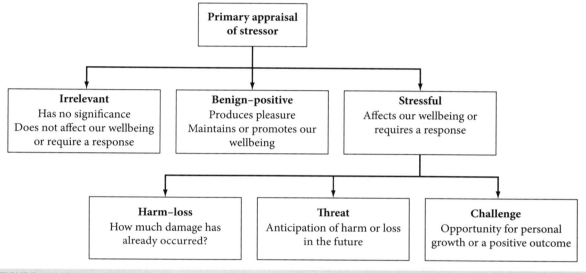

FIGURE 1.13 Primary appraisal

In **secondary appraisal**, an individual assesses their control over the situation and their coping resources and options. On the basis of this, the individual may then go back and reappraise the stressor and how they are going to respond.

These judgements play a fundamental role in determining not only the magnitude of the stress response, but also the kind of coping strategies that the individual may employ in efforts to deal with the stress. If a person believes that a challenge will severely tax or exceed their resources, they will experience stress and a corresponding physiological response.

Lazarus' model proposes that if stressors are perceived as positive or challenging rather than a threat, and if the person is confident that they possess adequate rather than deficient coping strategies, stress may not necessarily follow the presence of a potential stressor. Stress can therefore be reduced when stressed people are helped to change their perceptions of stressors, providing them with strategies to help them cope and improving their confidence in their ability to do so.

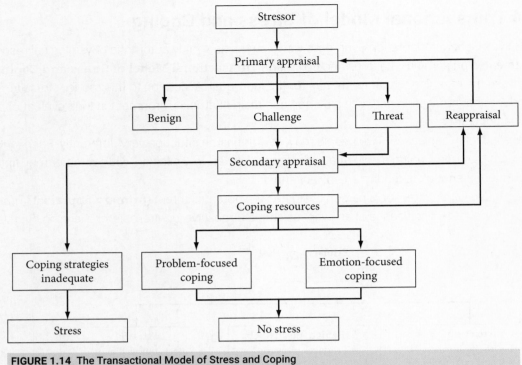

FIGURE 1.14 The Transactional Model of Stress and Coping

In their Transactional Model, Lazarus and Folkman distinguish different ways of dealing with stress.

Problem-focused coping involves taking practical action to directly tackle the problem or stressful situation that is causing stress, consequently reducing the stress by changing or eliminating its source. Problem-focused strategies include those that modify a person's behaviour, such as information-seeking, taking control, learning new skills and building up social supports and networks. In general, problem-focused coping is best as it removes the stressor, thereby dealing with the root cause of the problem to provide a long-term solution. These strategies best when the person has greater perceived control over their problem (e.g. exams, work-based stressors) and when the stressor is something that can be changed or modified.

However, it is not always possible or appropriate to use problem-focused strategies. Such approaches will not work in any situation where it is beyond the individual's ability to control or remove the source of stress. Appraisal-focused strategies occur when a person reassesses a situation, challenging their own assumptions to change their mindset or modify the way they think. Instead of tackling the cause, they change what they can by modifying their perception of what is happening and, as a result, alter their response to the situation.

Rather than changing the problem, **emotion-focused coping** is aimed at the effects of the stress, trying to target the thoughts and feelings associated with distress to reduce any negative emotional responses such as embarrassment, fear, anxiety, depression, excitement and frustration. This may be the only realistic option when the source of stress is outside the person's control. Some emotion-focused strategies, such as wishful thinking, denial or avoidance of the problem in the hope that it will go away, are maladaptive; they may make you feel better in the short term, but the problem will only come back and often be worse in the long term. Other strategies are more appropriate if the source of stress is unchangeable or outside the person's control and can facilitate expressing and processing emotions as a prelude to reappraising the stressor. Positive emotion-focused mechanisms, such as seeking social support, positive reappraisal and acceptance, are associated with beneficial outcomes.

1.2.5 Coping strategies

Stress management programs can only become effective if the individual's ability to eliminate, reduce or cope with stress is successfully assessed and changes are put into practice. **Coping strategies** refer to the behavioural and psychological efforts that people employ to master, tolerate or minimise their response to a stressor.

Context-specific effectiveness

Even though a coping strategy may be applicable and successful in one circumstance, it could be ineffectual or possibly detrimental in another. A coping strategy has **context-specific effectiveness** when it is an appropriate and adaptive choice to deal with a given stressor.

Factors affecting context-specific effectiveness include:
- contextual factors (situational dynamics; social and environmental influences)
- the type and nature of the stressor
- knowledge and understanding
- evaluation of the situation
- resources available
- access to social support from family, friends and community
- personal attributes, including skills and expertise
- general health/medical condition.

Coping flexibility

Not all coping mechanisms bring about positive coping; some types of coping mechanisms are maladaptive. The effectiveness of any one coping strategy and its impact on wellbeing may also vary from one context to another, generating different consequences that can make it effective or ineffective depending on the situation.

As we will have to contend with a diverse range of situations throughout our life, we need to be able to choose and implement suitable strategies to deal with each stressor we encounter.

Coping flexibility involves the ability to choose a coping strategy that is appropriate for the stressful situation, and successfully change or adapt it to meet the changing requirements of the situational dynamics. Should the strategy be inadequate to deal with the stressor, the individual should acknowledge this, cease the course of action and come up with a different approach. The ability to change and vary our methodology means that we are more able to deal with the wide assortment of challenges that we will encounter in a constantly changing world.

TABLE 1.6 Characteristics of individuals with differing levels of coping flexibility

Individuals with high coping flexibility	Individuals with low coping flexibility
• Tend to have a good understanding and application of context-specific effectiveness • Are likely to use a variety of coping strategies across different stressful situations • Quickly adapt the coping strategy if it is shown to be ineffective • Are more likely to achieve constructive results from the coping strategies they employ • Tend to cope better with stress	• Are not very adaptable or open to change • Always employ the same type of coping strategy despite differing situational dynamics • Continue to use a particular coping strategy, even if it is proving to be unproductive • Are less likely to benefit from the coping strategies they employ • Tend to not cope well with stress

Approach and avoidance coping strategies

The aim of coping strategies is to increase a person's ability to manage a situation in order to reduce stress levels. This can be done in a variety of ways. **Approach strategies** deal with problems directly, whereas **avoidance strategies** involve an individual pulling back from a situation or avoiding the problem. As the nature of stressors can be complicated, the individual must decide which type of strategy is going to be the most effective way to contend with the situation at that time.

TABLE 1.7 Comparison of approach and avoidance coping strategies

	Approach strategies	Avoidance strategies
Explanation	• Attention is focused on the stressor and its causes in order to devise a strategy to actively tackle the situation directly, consequently reducing the stress by changing or eliminating its source.	• Actions focus on detachment or withdrawal from the stressor, with no effort to tackle the situation or its causes (e.g. 'the elephant in the room').
Benefits	• Generally considered to be more adaptive and effective. • Necessary when action is needed to deal with a serious problem. • Usually associated with better outcomes, fewer psychological symptoms and more effective functioning. • Enable the person to look for ways of managing or removing the problem to provide a long-term solution.	• Avoidance strategies may enable short-term coping, especially if the stressor is too intense to deal with straight away. • Detachment might be appropriate when the source of stress is outside the person's control (such as waiting for the results of an exam). • If there are several issues to be dealt with concurrently, selectively avoiding certain stressors, especially if they are too complex, can enable the individual to focus on matters that can be resolved, helping them effectively respond to the larger problems at a later stage.
Limitations	• Difficulty coping with a number of stressors at once. • Ineffectual when the stressor is something that cannot be changed or modified. • Ineffective when the person does not feel that they have any control over their problem.	• Tend to be maladaptive. • Prolonged use can hinder people from dealing with stressors in productive ways. • A delay in dealing with a stressor can be harmful when a response is needed right away, as the problem will only come back and often be worse in the long term. • Excessive use is often linked with several damaging effects, such as an increased susceptibility to mental health problems and stress-related illness.
Comparative term in the Transactional Model of Stress and Coping	• Problem-focused coping	• Emotion-focused coping

Adaptive coping strategies generally involve confronting problems directly, making reasonably realistic appraisals of problems, recognising and changing unhealthy emotional reactions, and trying to prevent adverse effects on the body. Maladaptive coping includes using alcohol or drugs to escape problems.

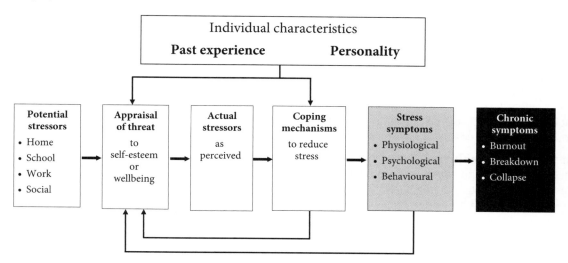

Potential stressors
- Family health or emotional problems
- Life crises
- Lack of social support
- Conflict between study/work and family demands
- Stressors intrinsic to study/school or to the job/workplace
- Career structure and/or options
- Relationships at school/work
- Financial difficulties

Coping mechanisms
- Exercise
- Relaxation (mental and physical)
- Social support
- Talking things over with someone
- Asking for help
- Being organised/planning ahead
- Setting priorities
- Balancing school/work and other interests/demands
- Diversity of interests
- Intellectual pursuits
- Taking a break/going on holidays
- Knowing your limitations
- A positive outlook
- A sense of humour

Stress symptoms
- **Physiological:** ulcers, high blood pressure, heart disease, infections, weight loss or gain, headaches/migraines, muscle tension, sore throat, back pain, skin problems (acne, eczema)
- **Psychological:** depression, increased anxiety, lowered self-esteem, sleep problems/nightmares, excessive tiredness, memory/concentration problems, withdrawing from social contact
- **Behavioural:** reduced interest in job/career, absenteeism and negative attitude, difficulty organising/prioritising tasks, lack of appetite, high emotionality, irritability, aggression, excessive drinking/smoking

FIGURE 1.15 A model of stress

Glossary

action potential (neural impulse) The electrical charge that travels down the axon of a neuron during transmission of a neural impulse, occurring as the result of the rapid depolarisation of the neuron's membrane, prompting the release of neurotransmitters

acute stressors Challenging situations that trigger an immediate response which is usually short lived, as the situations are dealt with and resolved quickly and normally have no long-term negative effects on health and wellbeing

adrenaline (epinephrine) A neurohormone that is released by the adrenal glands during the stress response, acting on the heart, lungs and muscles to optimise the body's fight–flight–freeze response to the stressor by increasing heart rate, oxygenation of blood and blood sugar levels, and relaxing smooth muscles to open airways

afferent pathway Ascending pathways within the somatic nervous system that transmit information from the sensory receptors to the brain via the spinal cord

agonist In neural communication, a substance that binds to a *neuroreceptor* to produce a similar effect to that of a *neurotransmitter* in either exciting or inhibiting a *postsynaptic neuron*; for example, *benzodiazepine* drugs are *GABA* receptor agonists that have sedative effects similar to GABA

alarm reaction stage The first stage of Selye's GAS response where resistance to stress is lowered and temporarily falls below normal, followed by a general defensive rebound (countershock), releasing adrenaline and cortisol to adapt to the stressor

antagonist In neural communication, a substance that suppresses the release of a neurotransmitter or blocks the receptor sites, making the postsynaptic neuron less likely to fire

anxiety An emotional state characterised by the anticipation of danger, dread or uneasiness as a response to an unclear or ambiguous threat

approach strategy Any response to managing stress (coping) that involves direct engagement with the stressor; consisting of behavioural or psychological responses designed to change (remove or diminish) the nature of the stressor and/or how a person thinks about it

autonomic nervous system (ANS) The component of the *peripheral nervous system* (PNS) that innervates involuntary (smooth) muscles and glands, including the organs of the circulatory, digestive, respiratory and reproductive systems; transmits information from the brain to organs and glands, and from these systems back to the brain

A+ DIGITAL FLASHCARDS Revise this topic's key terms and concepts by scanning the QR code or typing the URL into your browser.

https://get.ga/aplus-vce-psych-u34

avoidance strategy Any response to managing stress (coping) in which the person does not address the problem directly; diverting attention away from a threat or disengaging from a problem to escape painful or threatening thoughts, feelings, memories or sensations associated with the stressor

brain The master organ of the central nervous system within the skull, responsible for receiving and processing information from the rest of the body and generating responses to it

central nervous system (CNS) A major division of the nervous system consisting of all the nerves in the brain and spinal cord

challenge In the Transactional Model of Stress and Coping, the judgment when a situation is evaluated as placing a demand on our resources but viewed as an opportunity for personal growth or something that could lead to a positive outcome

chronic stressors Situations that are persistent and ongoing, leading to prolonged levels of physiological arousal that negatively affect health and wellbeing

conscious response Behaviours involving attention and awareness to make intentional decisions before sending commands via efferent neurons to enable a controlled response to a stimulus

context-specific effectiveness In relation to coping strategies, the effectiveness of a strategy is influenced by the degree to which it provides a good match to the situation (context)

coping flexibility The ability to stop an ineffective coping strategy (or evaluate the coping process) and implement an alternative effective coping strategy (or adapt the coping process)

coping strategy A deliberate action or thought process used to manage a stressful or unpleasant situation and/or to regulate one's response to such a situation

cortisol The primary stress hormone secreted into the bloodstream, enabling adaptive changes to energise the body in response to a stressor; however, prolonged abnormal levels in the bloodstream suppress the effectiveness of the immune system and can cause health problems

countershock The stage of alarm reaction in Selye's GAS, during which the sympathetic nervous system is activated and adrenaline increases, triggering a defensive fight, flight or freeze response

distress A negative psychological response to a stressor that results from being overwhelmed by the perceived demands of a situation, loss or threat

dopamine A modulatory neurotransmitter produced in the midbrain and adrenal glands that plays a major role in the coordination of movement and the regulation of reward; dopamine imbalance is associated with many neurological disorders (especially Parkinson's disease), and with many mental health problems, including addictive behaviours

efferent pathway Descending pathways that carry motor commands from the brain and spinal cord to the somatic nervous system to enable a response from organs, glands and muscles, such as voluntary and involuntary movements

emotion-focused coping In the Transactional Model of Stress and Coping, the process of dealing with ongoing stress by attempting to reduce any negative emotional responses, often by employing avoidance strategies

enteric nervous system (ENS) The largest component of the autonomic nervous system, which manages the functions of the digestive system (gastrointestinal tract). It can function independently of the CNS and so is sometimes called the 'second brain'. The extensive two-way neural connections between the ENS and CNS, particularly via the vagus nerve, form the gut–brain axis

eustress A positive psychological response to a stressor that has been appraised as a challenge rather than a threat; characterised by positive psychological and physiological responses that allow the person to meet the challenge effectively

excitatory neurotransmitter Neurotransmitters that increase the likelihood that a receiving neuron will fire an action potential

exhaustion stage The last of the three stages of Hans Selye's GAS, during which the adaptations made to resist the stressor break down, depleting the body of resources, causing symptoms such as increased risk of illness, sleep disturbance, fatigue and trembling

fight–flight–freeze response The body's automatic reaction to danger in which the autonomic nervous system mobilises energy and prepares the body for one of three responses: 1. confront the stressor (fight); 2. escape the stressor (flight); 3. immobilise to evade detection and prepare to respond if necessary (freeze). The fight and flight components are controlled by the sympathetic nervous system, whereas the freeze component is controlled by the parasympathetic nervous system. Physiological changes support heightened attention and fast/powerful physical responses through increased oxygenation of the bloodstream and diversion of energy from systems not needed for the response (e.g. digestion)

gamma-aminobutyric acid (GABA) The primary inhibitory neurotransmitter; its overall effects are to calm or slow neural transmission and therefore calm the body's response

gastrointestinal system Collective term for the organs involved in digestion, specifically the stomach and intestines, which contain the gut microbiota that play an important role in metabolism and are part of the gut–brain axis

General Adaptation Syndrome (GAS) The physiological model of the stress response proposed by Hans Selye that describes three stages of alarm reaction (shock/countershock), resistance (or adaptation) and exhaustion

glutamate The primary excitatory neurotransmitter in the brain, responsible for the fast transmission of neural messages and involved in cognitive functions

gut microbiota (or gut flora) The system of micro-organisms, including bacteria, that live in the gastrointestinal tract (digestive system), playing important roles in digestion and metabolism; also affect brain health and functioning through extensive connections between the enteric nervous system and central nervous system (gut–brain axis)

gut–brain axis (GBA) The network of bidirectional (two-way) communication pathways that allows communication between bacteria in the gastrointestinal tract (gut microbiota) and the brain, including communication via chemical transmission through the bloodstream, neuronal and hormonal pathways and via the immune system; causes disorders within the gut to affect the brain and vice versa

harm/loss In the Transactional Model of Stress and Coping, the judgment when a situation is evaluated as having already had a negative impact on our wellbeing

hippocampus A seahorse-shaped brain structure within the temporal lobes of the brain that is associated with memory formation, encoding declarative information and transferring it from short-term memory to long-term memory to form and consolidate new declarative explicit memories

homeostasis A state of internal equilibrium (balance), which an individual actively seeks to maintain

hypothalamic–pituitary–adrenal (HPA) axis A major part of the neuroendocrine system that regulates many bodily processes, in particular mediating the body's response to chronic, ongoing stress

hypothalamus A structure within the brain that triggers the fight–flight–freeze response through its link with the pituitary gland, leading to a state of increased arousal in reaction to a perceived threat

immune system The system in the body that mobilises bodily defences (such as white blood cells) against infection and disease by seeking out and destroying harmful influences such as viruses and bacteria

inhibitory neurotransmitter Neurotransmitters that decrease the likelihood that a receiving neuron will fire an action potential

interneuron Neurons within the CNS that relay messages from sensory neurons to other interneurons or to motor neurons; important in the integration of responses to peripheral information, such as the reflex arc

long-term depression (LTD) A form of neural plasticity that results in a long-lasting reduction in the strength of a neural response due to persistent weak stimulation

long-term potentiation (LTP) A form of neural plasticity that results in a long-lasting strengthening of neural connections at the synapse as a result of repeated stimulations from a presynaptic to postsynaptic neuron during learning. This improves the ability of two neurons to communicate with one another resulting in enhanced or more effective functioning of the neurons whenever they are activated. Demonstrated to occur in cells of the hippocampus, producing the neural changes that underlie the formation of memory

microbiota (or gut flora) The system of micro-organisms, including bacteria, that live in the gastrointestinal tract (digestive system), playing important roles in digestion and metabolism; also affect brain health and functioning through extensive connections between the enteric nervous system and CNS (gut–brain axis)

motor neurons Specialised efferent neurons in the CNS that carry motor commands from the brain and spinal cord to muscles, organs and glands to control voluntary and involuntary movements. Upper motor neurons carry information from the brain to the spinal cord; lower motor neurons form connections between the spinal cord and muscles, organs and glands

nervous system The integrated network of neurons, nerves, nerve tracts and associated organs and tissues, including the brain, which together coordinate a person's functioning, behaviours and responses adaptively as they interact with and adapt to their external environment

neuromodulator (modulating neurotransmitter) Neurotransmitters and hormones that act outside of the synapse to modify neuronal excitability, including dopamine in the reward system

neuron An individual nerve cell that receives, transmits and processes information

neuroplasticity (neural plasticity) The nervous system's ability to modify its structure and function as a result of experience and environmental stimulation (e.g. following an injury)

neurotransmitter Any chemical released from the axon terminal buttons of a presynaptic neuron into the synaptic cleft following an action potential that either excites or inhibits the postsynaptic neuron

noradrenaline (norepinephrine) A neurohormone produced in the brainstem and in the adrenal glands. It works together with epinephrine to support the stress response, constricting arteries to increase blood pressure, heart rate and blood sugar levels. As a neurotransmitter, it is released in response to emotional arousal and enhances the learning and memory for emotionally arousing events

parasympathetic nervous system The branch of the ANS that controls unconscious processes related to rest, repair and enjoyment, such as digestion, sleep, slowed heart rate, pupil constriction and sexual arousal, calming the effects of the sympathetic nervous system

peripheral nervous system (PNS) The division of the nervous system that comprises all of the nerves outside the CNS, through which motor information is communicated from the CNS to muscles and organs, and sensory information is communicated back to the CNS

primary appraisal In the Transactional Model of Stress and Coping, the first judgement we make when evaluating whether a situation poses a threat, harm or challenge; followed by secondary appraisal

problem-focused coping In the Transactional Model of Stress and Coping, the process of dealing with ongoing stress by taking practical action, employing approach strategies to directly tackle the situation that is causing stress, consequently reducing the stress by changing or eliminating its source

psychosomatic illness Actual physical symptoms and/or ailments that are attributed to a combination of physiological and psychological factors, such as emotional distress

receptors Cells located in the sense organs that are specialised to detect (receive) specific types of sensory information

rerouting In cases where damage has occurred within the brain, undamaged neurons may sprout new extensions to compensate for any loss of function. If an undamaged neuron has lost a communication point with an active neuron, it may seek a new active neuron and link with it instead, to allow messages to be sent along a new neural pathway to go around the damaged area

resistance stage The second of the three stages of Selye's GAS, during which the increases in cortisol and adrenaline and other physiological changes are stabilised to maintain the defensive response to the stressor; also called the adaptation stage

secondary appraisal In the Transactional Model of Stress and Coping, the judgement we make after primary appraisal to evaluate our ability and resources to cope with the stressor

sensory neurons Specialised afferent neurons (sensory receptors) located in sense organs that detect and respond to information (physical energy) from the environment and transmit it to the central nervous system

serotonin A modulatory neurotransmitter that regulates many functions, including mood, the sleep–wake cycle, memory processing and appetite

shock The first of stage of alarm reaction in Selye's GAS, during which the initial acute physiological response to the stressor occurs, causing a rapid drop in body temperature, blood pressure and muscle tone as though the body were injured

somatic nervous system The part of the PNS that transmits sensory information received from sensory receptors to the CNS, and motor messages from the CNS to skeletal muscles

spinal cord Part of the CNS consisting of a cable of sensory and motor nerve fibres that extends from the brainstem through a canal in the centre of the spine to the lumbar region of the spine; transmits sensory information from the PNS to the brain, and motor messages from the brain to the PNS

spinal reflexes The simplest kind of automatic, unlearned responses to stimuli (reflex) controlled by simple sensory-motor circuits in the spinal cord (bypassing the brain). Includes reflexes comprising a sensory and motor neuron connected by an interneuron (e.g. withdrawal reflex) and reflexes in which there is a direct connection between a sensory and motor neuron (e.g. patellar knee-jerk response). Also called simple reflex arcs

sprouting Following stimulation and repeated activation, the growth of more neuronal branches (dendrites and axon terminals) to make more connections (synapses) with other neurons

stress The physiological and psychological responses that a person experiences when confronted with a situation that is threatening or challenging

stressor The object, entity or event that causes a feeling of stress

sympathetic nervous system The branch of the ANS that alters the activity level of internal muscles, organs and glands to physically prepare our body for increased activity during times of high emotional or physical arousal

synapse The specialised junction between a presynaptic and postsynaptic neuron separated by a small gap called the synaptic cleft; enables neural signals to be transmitted when an action potential causes the presynaptic neuron to release a neurotransmitter into the synaptic cleft, where it can bind to receptors on the postsynaptic neuron

synaptic plasticity The ability of neurons to change their structure to create new synapses or eliminate redundant ones

synaptic pruning The elimination of redundant synaptic connections, particularly during the brain's early development

synaptogenesis The formation of new synapses as a result of learning, particularly during the brain's early development

threat In the Transactional Model of Stress and Coping, the judgment when a situation is evaluated as having the potential to negatively affect our future wellbeing

Transactional Model of Stress and Coping (Lazarus and Folkman Model of Stress and Coping) The theoretical model of stress and coping proposed by Lazarus and Folkman (1987) that emphasises the interaction (transaction) between a person and their environment; a person's response to stress is strongly influenced by how they appraise (judge, evaluate) the relevance of the stressor (primary appraisal) and their capacity to cope with it (secondary appraisal)

unconscious response Behaviours that do not involve our immediate, direct awareness in order to occur, such as reflexes

vagus nerve The tenth cranial nerve that is the major communication route in the gut–brain axis, extending from the brainstem to provide parasympathetic innervation to many organs, including the gut and digestive organs

Revision summary

Use the following summary of syllabus dot points and key knowledge within Unit 3 Area of Study 1 to ensure that you have reviewed the content thoroughly. Provide a brief definition or comment for each item to demonstrate your understanding or code them using the traffic light system: green (all good), amber (needs some review) or red (priority area to review). Alternatively, write a follow-up strategy.

How does the nervous system enable psychological functioning?	
Nervous system functioning	
• The role of the central nervous system	
• The role of the spinal cord	
• The role of the peripheral nervous system	
• The role of the somatic nervous system	
• Sensory pathways receiving and processing sensory stimuli	
• Motor pathways coordinating responses to internal and external stimuli	
• Conscious responses to stimuli	
• Unconscious responses to stimuli	
• Spinal reflexes	
• The role of the autonomic nervous system	
• The role of the sympathetic nervous system	

• The role of the parasympathetic nervous system	
• The role of neurotransmitters in the transmission of neural information across a neural synapse	
• Excitatory neurotransmitters	
– Glutamate	
• Inhibitory neurotransmitters	
– GABA (gamma-aminobutyric acid)	
• Synaptic plasticity	
• Changes to connections between neurons as the fundamental mechanisms of memory formation that leads to learning	
• Long-term potentiation	
• Long-term depression	
• Sprouting	
• Rerouting	
• Pruning	

Stress as an example of a psychobiological process	
• Stress	
• Stressors	
• Internal stressors	
• External stressors	
• Responses to stress	
• The fight–flight–freeze response in acute stress	
• The role of cortisol in chronic stress	
• The gut–brain axis (GBA)	
• The bidirectional communication between the central and enteric nervous systems	
• The relationship between cortisol, the composition of gut microbiota and stress and anxiety	
• Biological response to stress	

»	• Selye's General Adaptation Syndrome	
	– Alarm reaction (shock/countershock)	
	– Resistance	
	– Exhaustion	
	• Psychological response to stress	
	• Lazarus and Folkman's Transactional Model of Stress and Coping	
	– Primary appraisal	
	– Secondary appraisal	
	• Coping strategies	
	– Approach strategies	
	– Avoidance strategies	
	• Context-specific effectiveness	
	• Coping flexibility	

Exam practice

Nervous system functioning

Answers start on page 217.

Multiple-choice questions

Question 1

The two main branches of the nervous system are

A the brain and spinal cord.

B dendrites and axon terminals.

C the somatic division and autonomic division.

D the central nervous system and the peripheral nervous system.

Question 2

A major function of the spinal cord is to

A convey messages to and from the central nervous system.

B transmit messages to and from the brain.

C initiate voluntary movement.

D hold the body upright.

Question 3

The two main divisions of the peripheral nervous system are the

A afferent nervous system and the efferent nervous system.

B spinal cord and the somatic nervous system.

C autonomic nervous system and the somatic nervous system.

D sympathetic nervous system and the parasympathetic nervous system.

Question 4

As you read this question, your eyes are moving from word to word under the direction of the

A autonomic nervous system.

B parasympathetic nervous system.

C somatic nervous system.

D sympathetic nervous system.

Question 5

Which of the following is **not** a function of motor neuron activity?

A The transmission of messages from the central nervous system to the effector organs in the periphery of the body

B To carry commands from the central nervous system to the voluntary muscles

C To provide feedback from the peripheral nervous system to the central nervous system

D The transmission of signals to glandular organs in the body

Question 6

Which of the following processes is controlled by the autonomic nervous system?

A Automatic reflex reactions, such as when touching a hot kettle
B Becoming flushed when nervous
C Coordination of kinaesthetic bodily movement when playing sport
D Being able to name various parts of one's own body

Question 7 ©VCAA 2010 E1 SA Q11

Which of the following is true of the autonomic nervous system (ANS)?

A The ANS is a vital part of the central nervous system (CNS).
B It is impossible to consciously influence the functioning of the ANS.
C The ANS ensures that the constantly changing energy requirements of the body are met.
D The ANS relays messages between the CNS and the voluntary muscles that control our internal organs and glands.

Question 8

When walking alone through the city late one night, Mark turned the corner to see a large gang of youths directly in his path. Seeing the situation as potentially dangerous, he experienced an adrenaline rush, and thought he could feel his heart pounding in his chest. He also noticed other reactions, including a dry mouth and deeper breathing, and that he had broken out in a cold sweat.

The fight-or-flight response Mark experienced is

A directed by the somatic nervous system.
B consciously controlled by the central nervous system.
C dominated by the sympathetic nervous system in response to a stressful situation.
D influenced by the parasympathetic nervous system in response to a stressful situation.

Question 9 ©VCAA 2020 SA Q4

Jacob is a paramedic attending to a patient who has been unconscious for at least 30 minutes.

While Jacob is checking the patient's pulse, the patient suddenly wakes up and yells loudly, startling Jacob.

Which immediate physiological changes would Jacob experience when he is startled?

A Constriction of airways and bladder relaxation
B Increase in salivation and release of adrenaline
C Release of cortisol and increased release of glucose
D Increase in adrenaline and reduced movement in the large intestine

Question 10

While she was out walking, Aisha was startled by what she thought was a gunshot, causing her heart to race and her blood pressure to increase. She was relieved when she realised that it was only a car backfiring down the street. As she continued her walk, Aisha's blood pressure and heart rate returned to normal levels due to activity in her

A sympathetic nervous system.
B somatic nervous system.
C autonomic nervous system.
D parasympathetic nervous system.

Question 11 ©VCAA 2020 SA Q2

Which statement about conscious or unconscious responses by the nervous system is correct?

A A conscious response by the nervous system is involuntary and goal-directed.

B A conscious response by the nervous system is voluntary and attention is given to the stimulus.

C An unconscious response by the nervous system is voluntary and regulated by the autonomic nervous system.

D An unconscious response by the nervous system is unintentional and is always regulated by the autonomic nervous system.

Question 12

Rafael's family is building an extension onto the back of their house. While he was walking barefoot across their back lawn, he stepped on a nail. Instantly, he reacted by withdrawing his foot.

This response involved all the following parts of the nervous system **except**

A the brain.

B sensory neurons.

C interneurons.

D motor neurons.

Question 13

Which of the following statements is **not** true of neurotransmitters?

A Neurotransmitters are stored in synaptic vesicles.

B A neurotransmitter conveys a message to a neighbouring neuron by travelling along the axon to the terminal fibres.

C Neurotransmitters bind to postsynaptic receptor sites and subsequently trigger firing.

D When the neural impulse reaches the axon terminals, the vesicles release varying amounts of neurotransmitters.

Question 14

Neural impulses are passed along the axon until they reach the axon terminal. In order to pass the message to the next neuron, what must occur at the synapse?

A The electrical charge jumps from the axon terminal across the synaptic cleft to the dendrites of the next neuron.

B The axon terminals make contact with the dendrites of the next neuron and neurotransmitters are transferred.

C Synaptic vesicles in the axon terminal release neurotransmitters into the synaptic cleft, which lock into receptor sites of receiving dendrites.

D A change occurs in the postsynaptic neuron, but further research into the process is required because the exact mechanism is unknown.

Question 15

Neurotransmitters are chemical messengers that

A stay in the synapse rather than make contact with the next neuron.

B cause a change in the postsynaptic neuron by exciting the neuron and causing it to fire.

C cause a change the postsynaptic neuron by inhibiting the neuron and stopping it from firing.

D cause a change the postsynaptic neuron by either exciting or inhibiting the next neuron.

Question 16 ©VCAA 2017 SA Q4

Glutamate plays a key role in synaptic plasticity by

A releasing neurohormones into the bloodstream.

B increasing the speed of neurotransmitter transmissions along the axon.

C acting as an excitatory neurohormone released across the synaptic gap.

D acting as an excitatory neurotransmitter released across the synaptic gap.

Question 17

Inhibitory neurotransmitters act

A to make it easier for a postsynaptic neuron to generate an action potential.

B in conjunction with excitatory neurotransmitters to increase the likelihood that the postsynaptic neuron will be activated.

C to make it more difficult for the next neuron to fire.

D None of the alternatives is correct.

Question 18

Neuromodulators differ to neurotransmitters because they

A produce a slow and long-lasting effect.

B are rapidly reabsorbed by the presynaptic neuron through the process of reuptake.

C act directly by binding to specialised receptors on the postsynaptic neuron adjacent to their release point in order to cause a specific response.

D act on a specific synapse during direct synaptic transmission, thereby affecting one or two postsynaptic targets (neurons or effector cells).

Question 19

Psychological functions that involve dopamine include

A alertness and arousal, and awakening from the sleep cycle.

B voluntary movement, attention, motivation and pleasure/reward seeking.

C cognitive functioning associated with learning and consolidation of memory, particularly in the hippocampus.

D an inhibitory role, slowing down neural activity in the brain.

Question 20

The neuromodulator that plays a major role in mood regulation, sleep/wake cycles, memory processing and appetite is

A serotonin.

B dopamine.

C noradrenaline.

D glutamate.

Question 21

In researching the neural basis of memory, Kandel classically conditioned the *Aplysia* (a sea slug) by repeatedly giving it a mild electric shock after it was squirted with water, so that it learned to withdraw its gills whenever it was squirted.

Kandel's findings indicated that one of the fundamental mechanisms of memory formation is the process of long-term potentiation. This process involves

A a weakening in communication across synapses.

B a decrease in the release of neurotransmitters, especially glutamate.

C the strengthening of neural connections via repeated stimulation.

D increased levels of GABA within the synapse.

Question 22

To investigate whether environmental factors would affect the development of neurons in the cerebral cortex, Rosenzweig and his colleagues raised three groups of rats in different conditions. The first was a large group raised in an enriched condition in a large cage with several other rats and lots of stimuli. The second had three rats in a standard cage with food and water. The third was an impoverished condition, consisting of a lone rat with only basic necessities.

After 80 days, the rats were euthanised and their brains dissected. The researchers found that the first group of rats, reared in the enriched environment, had developed a thicker and heavier cerebral cortex than the rats in the other two groups.

On closer examination, this difference was attributed to

A a swelling of the neurons for the 'enriched' rats due to the increased release of the neurotransmitter acetylcholine.

B the 'enriched' rats developing larger cortical neurons that had sprouted longer and bushier dendrites to create more new synapses with other neurons.

C a sudden increase in postnatal proliferation of new neurons within the cerebral cortex of the 'enriched' rats.

D the neurons of the 'enriched' rats had fewer branches, allowing more room for existing dendrites to grow longer and connect with neurons that were further away.

Question 23

Which of the following may play a role in synaptic plasticity due to long-term potentiation?

A An increased release of glutamate from the axon terminal

B A decrease in the sensitivity of postsynaptic receptors

C Pruning of redundant synapses

D All of the above

Question 24

In relation to neural plasticity, long-term depression refers to

A a chronic mental disorder involving despondent moods.

B the weakening of neural connections due to habituation from excessive repeated stimulation.

C the process of increasing the number of glutamate receptors to compensate for a decrease in stimulation.

D selective weakening of specific synapses.

Short-answer questions

Question 25 (5 marks)

a   Describe the function of the central nervous system and identify its two main components.   2 marks

b   Outline the three functions of the spinal cord.   3 marks

Question 26 (5 marks)

a   Briefly describe the characteristics and function of the peripheral nervous system.   1 mark

b   Differentiate the main branches of the peripheral nervous system by outlining the role each one plays in controlling bodily functions.   4 marks

Question 27 (4 marks)

Outline the main function of each of the subdivisions of the autonomic nervous system listed below. Clarify your answer with some examples of physiological changes that would occur within the body as a result of their activity.

a   The sympathetic nervous system   2 marks

b   The parasympathetic nervous system   2 marks

Question 28 (7 marks)

A grazing gazelle is confronted by a cheetah in the wild. After a hectic chase, it is caught by the predator, and its apparently lifeless body is dragged to nearby bush. Knowing the cheetah is fully expended by the pursuit, nearby hyenas move in to try to take her prey, causing the cheetah to defend her catch. But the gazelle was only playing dead and uses the opportunity to run off when left alone by the cheetah.

a   With reference to the scenario, briefly describe how the fight–flight–freeze response enables a course of action to maximise the potential for an animal's survival.   2 marks

b   Summarise the role of the autonomic nervous system in the fight–flight–freeze response.   3 marks

c   Clearly explain the fight–flight–freeze response as an adaptive survival mechanism for any animal.   2 marks

Question 29 (7 marks)

While Meera was reading the morning paper, she reached for her coffee mug only to find it was empty. When she went to make another drink, she inadvertently touched the side of the kettle, instantly pulling her hand away before realising that the water had just been boiled.

a   Name the type of response and indicate the division of the nervous system involved when Meera reached for her coffee mug.   2 marks

b   i Identify the type of response involved when Meera pulled her hand away from the hot kettle.   1 mark

   ii Explain this response in terms of nervous system functioning. Ensure that you clarify how the brain was involved in this process.   4 marks

Question 30 (5 marks)

a   Explain the role of neurotransmitters in the transmission of neural information between neurons.   2 marks

b   Explain what is meant by the 'lock-and-key' analogy of neural transmission.   3 marks

Question 31 (4 marks)

a   Differentiate between excitatory and inhibitory neurotransmitters.   2 marks

b   Explain the role that glutamate plays in learning and memory.   2 marks

Question 32 (5 marks)

a Describe the role of neuromodulators within the nervous system. 2 marks

b Outline three differences between neurotransmitters and neuromodulators. 3 marks

Question 33 (9 marks)

Kandel's work with the *Aplysia* demonstrated that the formation of memory involved biochemical and structural changes within and between neurons in the brain.

a Briefly describe the neural mechanisms involved in learning. 1 mark

b Clarify the role of neurotransmitters in synaptic plasticity. 2 marks

c Describe the process and effect of long-term potentiation. 2 marks

d Outline the physiological changes that occur due to learning. 2 marks

e Describe what is involved in the process of long-term depression and its effect on synaptic plasticity. 2 marks

Question 34 (3 marks)

Clarify the role of the following processes within synaptic plasticity.

a Sprouting 1 mark

b Rerouting 1 mark

c Pruning 1 mark

Stress as an example of a psychobiological process

Answers start on page 223.

Multiple-choice questions

Question 1

A stress reaction

A is due to the action of the limbic system.

B occurs whenever a situation is perceived as a threat.

C happens when the parasympathetic nervous system is activated.

D depends on whether or not the somatic nervous system is involved.

Question 2

You are feeling stressed about your first Psychology SAC. As you wait for the task to begin, you become aware of certain physiological changes, such as dryness in your throat and mouth, a pounding heart and trembling. These reactions are caused by the

A release of adrenaline into the bloodstream.

B relaxation of the sympathetic nervous system.

C arousal of the parasympathetic nervous system.

D release of pituitary hormones into the bloodstream.

Question 3

The gut–brain axis describes

A how the gastrointestinal system is self-regulating and autonomous, essentially having its own brain.

B ascending afferent pathways carrying feedback from the intestines to the hypothalamus in order to control hunger.

C how descending efferent pathways from the brain send messages to the gut to monitor and integrate autonomic gut functions.

D the bidirectional communication network between the central and enteric nervous systems.

Question 4

As a part of the gut–brain axis, the gastrointestinal microbiome

A produces 95% of the body's supply of the neurotransmitter dopamine.

B digests food to supply energy and nutrients to the body.

C interacts directly with the CNS by influencing brain chemistry and affecting neuroendocrine systems associated with the stress response, anxiety and memory function.

D maintains a stable homeostatic environment that is unaffected by external factors, such as when diet varies or stress levels change.

Question 5 ©VCAA 2020 SA Q10

Which one of the following is the most accurate description of the role of cortisol in the stress response, according to Selye's General Adaptation Syndrome?

A Stops the immune system from functioning

B Increases glucose in the bloodstream and reduces inflammation

C Reactivates functions that are non-essential in a fight–flight response

D Provides the initial alert about a perceived threat, through the release of adrenaline

Question 6 ©VCAA 2019 SA Q9

Jamie is experiencing a constant state of stress and has also caught a cold.

Which of the following most accurately identifies the stage of Selye's General Adaptation Syndrome that Jamie is in and the reason that supports this stage?

	Stage	Reason
A	Shock	Jamie's immune system is immobilised so his body can fight the stressor.
B	Resistance	Continued cortisol release weakens Jamie's immune system, resulting in his body being unable to fight the cold.
C	Exhaustion	Jamie's body's resources are depleted, resulting in vulnerability to a range of serious physical disorders.
D	Resistance	Increased adrenaline in Jamie's bloodstream results in his body becoming susceptible to illnesses.

Question 7

Out of the blue, Wei's boss told him that his casual position no longer existed because of restructuring and he would need to find another job. Wei felt faint and weak at the knees, not knowing what he was going to do. Despite experiencing initial panic, Wei decided to channel his efforts into finding a new and better part-time position.

During the alarm reaction phase of the General Adaptation Syndrome, Wei's _____ nervous system would have been particularly active.

A central

B parasympathetic

C somatic

D autonomic

Question 8

To study her desired tertiary course, Hoana had to move out of her parents' home near Echuca and take up residence at a campus in Melbourne. The city lifestyle was a lot faster and noisier than life on the farm, and Hoana found that she had to do a lot of adjusting to fit in.

Her coursework was more demanding than her VCE studies the previous year and, in addition to this, she had to fit in a part-time job to help her financial situation.

In the early part of the year, Hoana complained of headaches and of always feeling tense and tired. Despite little changing (if anything, her workload is increasing), she now says that she feels 'okay' and is currently meeting the deadlines for her work.

Hoana is **most** probably

A successfully adjusted to her new lifestyle.

B at risk of developing a mental disorder within the exhaustion stage of the General Adaptation Syndrome.

C in the resistance stage of the General Adaptation Syndrome.

D denying the existence of any problems.

Question 9

A person's bodily resources are depleted and their stress hormones, adrenaline and noradrenaline, have run out. Without some form of stress management, severe illness may occur.

During which stage of Hans Selye's General Adaptation Syndrome would serious health problems occur?

A The stage of exhaustion

B The stage of resistance

C The shock phase within the alarm reaction stage

D The normal level of resistance stage

Question 10

Which of the following statements most accurately illustrates the relationship between stress and disease?

A Research has established a definitive cause-and-effect relationship between prolonged stress and heart disease, stomach ulcers and cancer.

B Research has shown that when undergoing prolonged stress, the immune system may be suppressed, resulting in harmful cells not being detected and therefore not eliminated quickly enough, which may result in illness.

C Research has shown that when undergoing prolonged stress, the sympathetic nervous system is activated, causing an increase in the immune system's functioning, which aids the body to detect and eliminate harmful cells effectively.

D In the immune system, lymphocytes and phagocytes are overactivated in a stressful situation, leading to psychosomatic illnesses such as asthma, ulcers and migraines.

Question 11

Since his wife stopped working due to ill heath, Hao has been working two jobs in order to pay the bills for his family.

A common physiological effect of the prolonged arousal that Hao may be experiencing could be _____, while a common psychological effect of prolonged arousal is _____.

A anxiety; heart disease

B stomach ulcers; aggression

C forgetfulness; fatigue

D headaches; eczema

Question 12

Yusuf and Ajay are asked to play their guitars before a school assembly. Yusuf is thrilled by the idea and looking forward to it, but Ajay is on the verge of a panic attack.

Why do different people react to the same stressor in different ways?

A Their past experiences of stress are different.

B They have different strategies to help them cope with stressors.

C They may appraise various stressors in different ways.

D All of the alternatives are correct.

Question 13 ©VCAA 2012 E2 SA Q24

The night before a university examination, Terri realised that there was an entire topic that she had not studied. She burst into tears and was unable to sleep that night.

According to Lazarus and Folkman's Transactional Model of Stress and Coping, it would be most likely that Terri's appraisal of this experience was

A a threat.

B eustress.

C irrelevant.

D a challenge.

Question 14 ©VCAA 2014 SA Q38

In Lazarus and Folkman's Transactional Model of Stress and Coping, the difference between a primary appraisal and a secondary appraisal is the

A primary appraisal is always conscious, whereas the secondary appraisal is not.

B secondary appraisal is always conscious, whereas the primary appraisal is not.

C primary appraisal evaluates the situation, whereas the secondary appraisal evaluates resources for coping.

D primary appraisal involves problem-focused coping strategies, whereas the secondary appraisal involves emotion-focused coping strategies.

Question 15

One advantage of Lazarus and Folkman's Transactional Model of Stress and Coping is that it

A is simple to test through objective experimental research.

B highlights physiological responses to the stressor.

C does not take cognitive evaluations into account when analysing the stress response.

D emphasises a person's psychological response to the stressor.

Question 16

A maladaptive method of coping would be

A denial.

B cognitive reappraisal.

C problem-focused approaches.

D stress management techniques.

Question 17 ©VCAA 2017 SA Q23

Rose participated in a television game show about general knowledge. While waiting to go onstage she felt very stressed. At one point she felt so anxious that she left the studio. However, she then decided to return to the studio and study the notes she had prepared earlier.

The coping strategies that Rose used prior to going onstage can be considered to be

A approach strategies only.

B avoidance strategies only.

C exercise and approach strategies.

D avoidance and approach strategies.

Question 18 ©VCAA 2020 SA Q45

Theodore lost his job two years before he intended to retire.

Theodore's unemployment was making him feel stressed. He applied for jobs advertised in the newspaper, but this did not result in employment. He went to see a careers counsellor, who suggested some further strategies for finding new employment.

Which one of the following statements applies to the coping strategy used by Theodore when he saw a careers counsellor?

A It is an approach strategy that does not demonstrate coping flexibility.

B It has context-specific effectiveness and demonstrates coping flexibility.

C It has context-specific effectiveness and will help him avoid stressful situations.

D Given the difficulty older people have finding a job, it demonstrates coping inflexibility.

Short-answer questions

Question 19 (6 marks)

a Using examples of each to clarify your answer, explain the difference between stress and stressors. 4 marks

b Identify the major pathway by which the brain may send signals to the endocrine system in times of the following.

  i Acute stress 1 mark

  ii Chronic stress 1 mark

Question 20 (4 marks) ©VCAA 2015 SB Q12

Although she is able to function in her everyday life, Annie is stressed about her driving test scheduled for today.

With reference to the physiological aspects of Annie's stress response, give two reasons why this level of stress may be helpful when Annie takes her driving test.

Question 21 (8 marks)

a Outline what is meant by the term 'gut–brain axis' (GBA). Clarify the relationship between the key elements of this axis.   3 marks

b Summarise the effect gut microbiota can have on the nervous system.   2 marks

c Will is experiencing some intense stress because of events both at home and at school. Consequently, he is now suffering from bad indigestion. Explain why Will's digestive system has been affected by the intense stress.   3 marks

Question 22 (13 marks)

According to Hans Selye, when an organism is exposed to an ongoing stressor, the General Adaptation Syndrome (GAS) occurs over three stages.

a Using appropriate terminology from Selye's model, outline how a person would respond to a stressor within the first stage of the GAS.   3 marks

b If stress continues, the person progresses to the second stage of the GAS. Using appropriate terminology from Selye's model, describe what happens during this stage.   2 marks

c Should the stress be prolonged, the person will enter the final stage of the GAS. Summarise what you would expect to happen during this stage.   2 marks

d At the beginning of the year, Year 12 students are often given lists of dates and deadlines for all of their assessment tasks and continue working throughout the year to prepare for their examinations. It is not unusual for students to get sick during Term 3, especially during or just after the peak periods for work requirements and SACs. Apply your understanding of Selye's General Adaptation Syndrome to explain this trend.   6 marks

Question 23 (3 marks) ©VCAA 2017 SB Q1 (ADAPTED)

Name each stage of Selye's General Adaptation Syndrome as it applies to each role of cortisol given in the table below.

	Role of cortisol	Name of stage
a	Sustained levels of cortisol mobilise the body and increase arousal to respond to the stressor.	
b	The release of cortisol mobilises the body and increases arousal to respond to the stressor.	
c	Depleted levels of cortisol reduce the ability of the body to respond to further stressors.	

Question 24 (6 marks)

Avantiika aspires to be a doctor and so she strives to achieve perfect results in all her SACs. She is also working part-time on weekends to save money to put towards the fees associated with the university course.

Lately, Avantiika has been continually catching colds and needing to take days off to recover. She is concerned about taking time off because she is so busy; a day off means that she gets further behind in her work. Avantiika often finds herself doing extra study late at night in order to catch up and meet deadlines. As a result, she is always feeling tired and lacking energy, and is constantly getting sick.

a From the information provided, identify and explain which stage of the GAS Avantiika is probably in.   2 marks

b If a person is subjected to chronic excessive stress, the first signs of illness are likely to appear during which stage in the General Adaptation Syndrome?   1 mark

c Explain in physiological terms why Avantiika's state of being run-down has made her prone to catching colds.   3 marks

Question 25 (15 marks)

The Year 12 coordinator wanted to acknowledge students' achievements at the inter-school sports competition, so asked three of the students to speak at the next assembly. When asked, Carla was excited by the prospect and Ryan didn't give it a second thought, but Jiang froze, went pale and instantly broke into a cold sweat.

a Provide a brief overview of Lazarus and Folkman's Transactional Model of Stress and Coping. 2 marks

b With reference to the Transactional Model of Stress and Coping, explain why people don't experience the same type of stress in response to the same situation. 2 marks

c Outline the process of primary appraisal within the Transactional Model of Stress and Coping, clarifying the different ways a stressor may be perceived. 5 marks

d Using terminology from the Transactional Model of Stress and Coping, clarify the difference between eustress and distress. 4 marks

e Describe the function of secondary appraisal within the Transactional Model of Stress and Coping. 2 marks

Question 26 (4 marks) ©VCAA 2015 SB Q4

Identify **two** strengths and **two** limitations of Lazarus and Folkman's Transactional Model of Stress and Coping. 4 marks

Question 27 (5 marks)

a Define 'coping strategies'. 1 mark

b Explain the concept of context-specific effectiveness in relation to coping strategies. 2 marks

c Describe what is involved in coping flexibility. 2 marks

Question 28 (10 marks)

a Differentiate between approach strategies and avoidance strategies for dealing with stress. 2 marks

b Identify two benefits of applying approach strategies for dealing with stress and two limitations of this method. 4 marks

c Outline two benefits of applying avoidance strategies for dealing with stress and two limitations of this approach. 4 marks

Chapter 2
Area of Study 2: How do people learn and remember?

Area of Study summary

This area of study examines the interdependent processes of learning and memory by considering different models that have been proposed to explain the mechanisms involved. The distinction between learned and unlearned behaviours is clarified to enable analysis of learning theories and their application within a variety of contexts.

Building on previous knowledge about synaptic plasticity, the physiological basis of memory is explored by looking at how brain areas work together to encode, store and retrieve information. Further to this, information processing models are examined to look at processes that can affect our memory, including mnemonics that can enhance its function.

Area of Study 2 Outcome 2

On completing this outcome you should be able to:

- apply different approaches to explain learning to familiar and novel contexts
- discuss memory as a psychobiological process.

The key science skills demonstrated in this outcome are:

- analyse and evaluate data and investigation methods
- construct evidence-based arguments and draw conclusions
- analyse, evaluate and communicate scientific ideas.

Adapted from *VCE Psychology Study Design (2023–2027)* © copyright 2022, Victorian Curriculum and Assessment Authority

UNIT 3 / CHAPTER 2 Area of Study 2: How do people learn and remember? 49

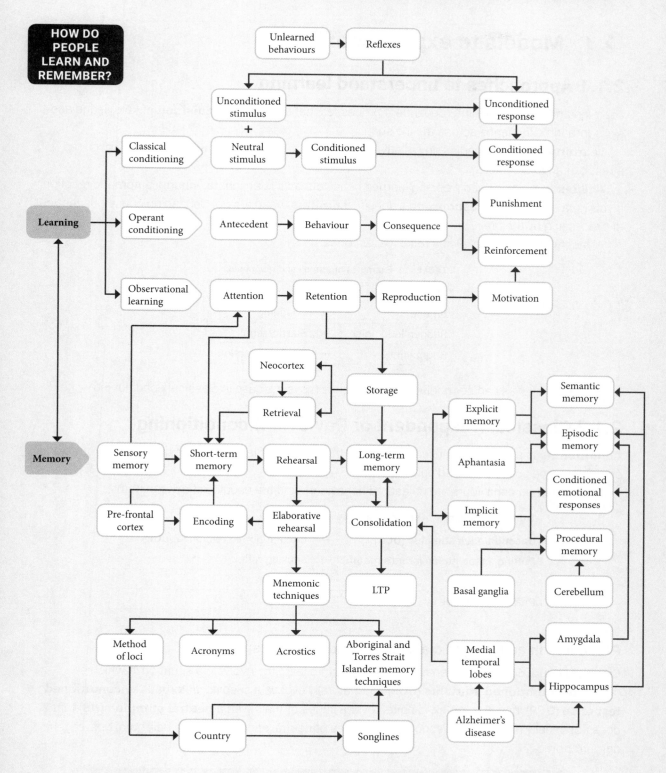

2.1 Models to explain learning

2.1.1 Approaches to understand learning

Learning and memory are interdependent processes that enable the **acquisition** of knowledge and skills through experience across the life span.

Learning can be defined as any relatively permanent change in **behaviour** that may occur as a result of an experience.

A **reflex** is not classified as being learned behaviour as it is an innate (inborn), automatic reaction, involving a simple, rapid **response** to a specific **stimulus** that does not depend on conscious thought or processing by the brain.

A variety of stimuli can elicit a reflexive response.

TABLE 2.1 Examples of stimuli and responses

Stimulus	→	Response
Smoke	→	Coughing
Sudden loud noise	→	Startle/fright
Bright light	→	Pupil contraction

However, reflexes can be modified by experience (as is the case in classical conditioning).

2.1.2 Classical (respondent or Pavlovian) conditioning

Conditioning is a term that is used to describe the process of learning when behaviours, events and stimuli become associated with each other.

Unconditioned behaviours are unlearned behaviours and are usually reflexive in nature.

TABLE 2.2 Examples of unconditioned behaviours

Unconditioned stimulus (UCS)	→	Unconditioned response (UCR)
Putting a teat into an infant's mouth	→	Sucking reflex
Seeing food	→	Salivation
Puff of air into the eye	→	Blinking

Procedure in acquiring a classically conditioned response

Classical conditioning is a behaviourist approach to learning involving a situation in which a certain **unconditioned stimulus (UCS)** that naturally evokes a specific, automatic **unconditioned response (UCR)**, usually a reflex, is paired over a series of trials with a **neutral stimulus (NS)** that does not usually produce this response. Examples of neutral stimuli include a bell, horn, buzzer, tone or metronome.

> **Note**
> Classical conditioning is also known by other names. To remember the different terms, use a different association for the abbreviation **CPR** (**C**lassical, **P**avlovian, **R**espondent).
> Alternatively, incorporate the initial letters into an acronym that you find easy to remember; for example, **CARP** (**C**lassical, **A**ssociative, **R**espondent, **P**avlovian) or **PRAC**.

TABLE 2.3 The three-phase model of classical conditioning

Phase 1	Before conditioning	Phase 1 can also be referred to as the 'baseline' or pre-conditioning condition. At this point, the UCS triggers the reflexive UCR, whereas the NS does not.
Phase 2	During conditioning	Phase 2 involves the process of 'acquisition' and can also be referred to as the 'conditioning phase', 'learning' or 'training'. During this stage, repeated pairings form an **association** or connection between the UCS and the NS. In order to become associated with one another, the NS should immediately precede the UCS and the two stimuli should be presented near each other in time and space to strengthen the association.
Phase 3	After conditioning	Phase 3 occurs post-conditioning and could be referred to as the 'testing' phase. In this phase, the learned behaviour has been clearly established when the once NS, now referred to as the **conditioned stimulus (CS)**, will elicit the **conditioned response (CR)** when presented alone, which it formerly did not produce.

Pavlov's experiments

Pavlovian conditioning is another term for classical conditioning. This term refers to the work of Ivan Pavlov (1849–1936), whose experiments with dogs highlighted the stages and processes involved in this form of conditioning.

Pavlov was initially researching the physiology of digestion but noticed that the salivation reflex in his dogs was occurring before the presentation of food at the sight of the lab attendant at the food storage cabinet or at the sound of his footsteps. Pavlov erected screens to prevent the dogs from seeing the food preparation. In addition, he tried using several stimuli to distract the dogs, but in each case the dogs learned to salivate to these stimuli as well. Curious, Pavlov then set up the series of experiments in the 1920s for which he is better known.

In order to control as many extraneous (and potentially confounding) variables as possible, Pavlov went about his research in the following way.

- The dogs were placed in an isolated, soundproofed room to eliminate any other possible stimuli.
- They were individually restrained in a harness to restrict their movement during the experiment.
- They were placed in a specially designed apparatus that enabled the mechanical positioning of meat powder directly in their mouths, or in a small bowl nearby without any human intervention.
- A surgically attached tube was inserted in the dogs' cheeks, which drained their saliva to then be measured by the experimenters.
- In the baseline stage, the dogs were initially presented with the sound of a bell (NS), which elicited no response.
- During the acquisition stage of conditioning, the bell was immediately rung before the dog received the meat powder (UCS), which naturally led to salivation (UCR).
- After several trials in which the bell and the food were associated, as a result of their **contiguity** in time and space, the acquisition stage was completed. The bell then became the CS, which led to a CR, salivation.

While descriptions usually focus on his use of a bell, Pavlov repeated his experiment procedure using a variety of different stimuli, leading to a CR in each case.

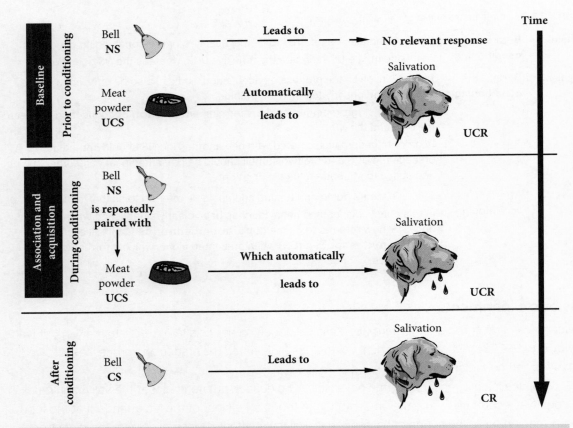

FIGURE 2.1 The classical conditioning procedure in Ivan Pavlov's (1920s) experiments on dogs

Factors that may influence whether a classically conditioned response can occur

The UCR needs to be an involuntary response over which the individual cannot exert control. Similarly, the UCS should be a stimulus that evokes a response with no requisite learning.

In order for a paired association to occur, a contiguous approach is necessary. This means that the NS and the UCS must occur close together in time and space in order for them to become linked. In animals and young children, this time factor should ideally be no more than half a second. Longer time periods are less effective in forming CRs.

The meaning of each term becomes clearer if broken down into parts.

- *Neutral = no effect/no relevant reaction*
- *Conditioned = learned*
- *Unconditioned = unlearned/innate*
- *Stimulus = something that evokes/triggers a response*
- *Response = reaction/behaviour*

Elements in classical conditioning

When two things occur together, we associate one with the other, so that when one appears, we expect the other to follow and react accordingly. Within classical conditioning, an **association** is formed between two stimuli, so that both will trigger the same response.

Stimulus discrimination is the ability to perceive the difference between two or more stimuli, even if they are similar in nature. Discrimination occurs in classical conditioning when a person or animal only shows the CR to the CS, but not to any other stimulus that is similar in nature. For example, Pavlov's dogs, through only being reinforced with food on trials with the CS, may learn to differentiate between the original bell (CS) and other new similar bell sounds.

Stimulus generalisation involves the likelihood that stimuli that are similar in nature will elicit the same response. Generalisation occurs in classical conditioning when a stimulus resembling the CS produces the same or a similar CR. For example, Pavlov's dogs may salivate (CR) after hearing buzzers or bells of a different pitch to the CS.

Extinction entails the termination of a CR. In classical conditioning, extinction occurs when the reinforcer, in the form of the UCS, is removed so that the association is eroded or broken. The CS loses its strength and eventually fails to elicit its CR.

After extinction has occurred, **spontaneous recovery** may be demonstrated by the reappearance of a CR. In other words, despite the prior weakening of the association due to the extinction process, the CR could recur when the CS is presented.

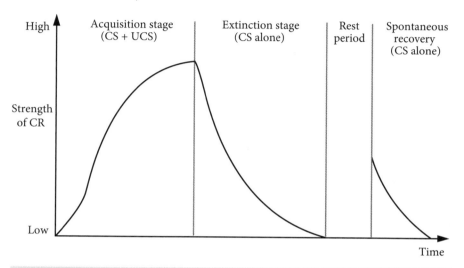

FIGURE 2.2 Extinction and spontaneous recovery

Applications of classical conditioning

Graduated exposure

Graduated exposure therapy, also known as **systematic desensitisation**, is mainly used to treat phobias, but can also be used for a number of other CRs, including taste aversions.

In the case of phobias, the clients are initially taught relaxation techniques that they can use at each stage of the process. They are then told to visualise the feared CS until they are able to do so in a relaxed manner. They are then taken, step by step, through a series of tasks (graded in order of difficulty from 0 to 100), where they learn to cope with and overcome the fear in each step of the hierarchy until the phobia is eliminated.

The process aims to gradually extinguish the CR or to change the CS that produced fear to become associated with a relaxed response.

> **Note**
> Systematic desensitisation will be explored along with phobias in Chapter 4.

Flooding

Similar to desensitisation, **flooding** is a therapeutic technique, generally used to treat phobias, which exposes a patient to vast amounts of the feared CS in order to hasten the process of extinction. While the patient would initially be very anxious, they begin to calm down when they realise that nothing bad will happen, enabling them to associate a feeling of calm with the previously feared object.

Flooding is not appropriate for every case, as it can trigger a higher level of sensitisation and fear reinforcement. The therapist would have to discuss with the patient the levels of anxiety they are prepared to endure during the process. A patient therefore needs to be highly motivated and be given appropriate support throughout the process.

Aversion therapy

Aversion therapy is a technique that uses classical conditioning principles, in which an individual learns to associate an undesirable learned behaviour pattern, such as gambling, aggression or illicit drug use, with an unpleasant response, such as fear or nausea.

For example, a psychologist may help someone quit smoking by showing them a diseased lung or blocked arteries whenever they have a puff on a cigarette, to produce disgust or revulsion. Other aversive stimuli could include unpleasant-tasting additives or even the use (with permission) of mild electric shocks. After repeated pairings, whenever the patient craves a cigarette, they will experience the negative response and therefore avoid smoking.

While some therapists may argue the ends as justifying the means, they must still adhere to ethical guidelines concerned with doing no harm.

Animal training

A similar approach to aversion therapy could be used in animal training in order to condition the animal to avoid certain hazards, such as wildlife or toxic plants. The aim of the process would be to pair an unpleasant UCS, such as a low-level electric shock or an unpleasant smell, with an NS, being the environmental hazard, in order to establish an avoidance response to that stimulus. Stimuli that cause a startle or pain response are better to use as UCS in this type of scenario because they are strongly related to innate survival mechanisms, which lead to faster learning.

Advertising

Advertisers often use the principles of classical conditioning when promoting products. The product (NS) is paired with a particular person, object or event (UCS) that is expected to trigger *positive emotions* (UCR) in the consumer. **Advertisements** are played repetitively to reinforce the association between the two stimuli with the intent of making potential customers experience positive emotions (CR) when thinking about the product (now CS) to make it more likely for the consumer to purchase the product.

2.1.3 Operant (instrumental) conditioning

The foundations for operant conditioning were established by Edward Thorndike (1874–1949) in his laboratory investigations of animal intelligence involving the behaviour shown by cats placed in a 'puzzle box'.

Based on his observations, Thorndike theorised that the cats learned to escape the puzzle box by trial and error. That is, they performed various responses in a blind, mechanical way, eliminating responses that did not achieve the desired outcome until some action was successful in leading to the desired outcome.

Thorndike concluded that the pattern demonstrated that the cats were instrumental in the learning process and in making the connection between their behaviour and its **consequences**. He postulated the **law of effect** to account for the behaviour of the cats. According to the law of effect, responses that are followed by a satisfying state of affairs will be repeated with greater and greater frequency over time. On the other hand, responses that are followed by an annoying or undesirable state of affairs will occur less frequently over time. In addition to this, his *law of exercise* proposed that repetition strengthens learning, making a behaviour easier to perform; in other words, 'practice makes perfect'.

Through his ongoing experiments, B.F. Skinner (1904–1990) developed the work of Thorndike into the theory of operant conditioning.

Operant conditioning is a behaviourist approach to learning where the consequences that follow a response determine whether the response is likely to be repeated.

Instrumental conditioning is an alternative name for operant conditioning, proposed by Thorndike to refer to the process in which an individual makes an association between behaviour and the consequences that follow it.

In order to explore a variety of factors influencing behaviour, and as responses can occur repeatedly and at any time, Skinner developed the operant chamber (now referred to as the 'Skinner box'). This type of apparatus can be automated, allowing constant monitoring of the experimental subjects (usually small animals).

Because the animal stays in the box, this apparatus can also run many trials in a row, allowing the manipulation of the conditions to vary reinforcement. As a result, a great deal of data can be collected, representing easily quantifiable response rates that demonstrate learning for each condition.

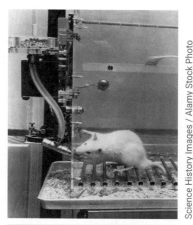

FIGURE 2.3 The Skinner box

TABLE 2.4 The three-phase model of operant conditioning

Phase		Description
1	**A**ntecedents (stimulus)	Phase 1 is referred to as the **antecedent** condition, the 'discriminative stimulus' or simply the 'stimulus'. In this phase, the learner is presented with a signal (stimulus) to associate with the production of the desired behaviour.
2	**B**ehaviour (response)	Phase 2 is referred to as the 'operant response' to the stimulus, or the 'behaviour'. In this phase, the desired behaviour is produced in response to the stimulus.
3	**C**onsequences (result)	Phase 3 is referred to as the 'consequence'. In this phase, the learner is rewarded when the desired behaviour is produced, or punished for undesirable responses.

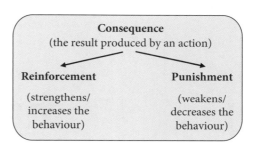

FIGURE 2.4 The consequence phase of operant conditioning

Reinforcement

Any object or event administered after a response that strengthens and increases the likelihood of that response reoccurring over time is said to be a **reinforcer**. This process is called **reinforcement**.

> **Note**
> When you *reinforce* something, you *strengthen* it so that it will last longer (e.g. reinforced concrete or folder paper).

Factors that may impact the effectiveness of reinforcement

The reinforcer should be administered as soon as possible after the desirable behaviour has been demonstrated to establish an association between the desired response and the reinforcer without other factors intervening. Furthermore, it may motivate the learner to progress.

Primary reinforcers are stimuli that normally lead to the satisfaction of innate, biological or physiological needs; for example, food, water, basic comforts and sleep.

Secondary reinforcers, on the other hand, are stimuli that normally lead to the satisfaction of needs through a learned association with a primary reinforcer; for example, money, good grades and success.

Positive reinforcement

Positive reinforcement occurs when a response is followed by a positive event that increases the likelihood of the response occurring again; for example, rewards, food, good grades, approval, compliments or money.

The positive reinforcer should have a high incentive value for the intended recipient or else it will be ineffective. For example, it is futile to offer a milkshake as a positive reinforcer to a child who is lactose intolerant.

The positive reinforcer should always be administered as a consequence of good behaviour and should not precede the good behaviour.

Negative reinforcement

Negative reinforcement occurs when a response is followed by an end to discomfort or by the removal of an unpleasant event or threat of punishment. There is an increase in the likelihood of the desired behaviour being strengthened over time. Examples include a child cleaning his room to end his mother's nagging, a student following school rules to avoid getting detention or a driver staying within the speed limit to avoid a fine or an accident.

> **Note**
> Negative reinforcement should not be confused with punishment, which aims to decrease the target behaviour.

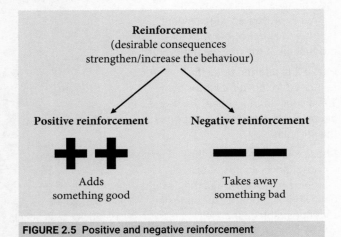

FIGURE 2.5 Positive and negative reinforcement

> **Note**
> *Positive* = something good and is associated with the symbol + (which also means addition in maths). Combining these two concepts explains the consequence in positive reinforcement. The converse is true for negative reinforcement.

> **Note**
> In the context of the consequences applied in operant conditioning, 'positive' refers to the *addition* of something and 'negative' refers to the *removal* of something – *not* 'good' and 'bad' as would be the more common meaning of these words. Hence the source of confusion associated with negative reinforcement.

The processes of positive and negative reinforcement can be distinguished in the following scenario. While at the supermarket, Beth was continually being pestered by her young son, Jack, to get a lolly. Losing her patience, she gave in and got him a lollipop, which then kept him quiet, allowing her to concentrate on her shopping. Next time they went grocery shopping, Jack repeated his nagging, so Beth gave him another lollipop because she knew this would stop his behaviour. This then became a behaviour pattern for successive shopping trips. Jack's behaviour, although undesirable, was *positively reinforced* by the lollies, whereas Beth's behaviour was *negatively reinforced* by the cessation of her son's whining.

Escape conditioning is based on negative reinforcement. It occurs when an individual learns to perform a response in order to end an unpleasant stimulus. For example, an individual confronted with an aggressive, pushy person may escape from a conversation with that person by making up the excuse that he is urgently needed elsewhere.

Avoidance conditioning involves both classical and operant conditioning and refers to the process of learning to make a response in order to evade discomfort. For example, an individual may walk away when they see a person they dislike approaching, which prevents an interaction from occurring. The individual first learns through classical conditioning that the person causes discomfort. Once this association is established, they begin to avoid the other person, which is likely to be repeated due to negative reinforcement, as it is followed by a sense of relief that they have managed to sidestep an unpleasant situation.

Punishment

In operant conditioning, **punishment** refers to any aversive or unpleasant consequence (*punisher*) for an individual that weakens or decreases the probability of a response recurring.

Positive punishment

Positive punishment involves the application of an undesirable outcome that is intended to decrease the likelihood of a response recurring; for example, smacking, electric shock or hard labour.

Negative punishment (response cost)

Response cost, also known as **negative punishment**, is a form of punishment involving the removal of a desirable event or commodity that decreases the likelihood of a response recurring. Examples include losing privileges for not doing what you were told, jail (taking away your freedom) for committing a crime, or a driver receiving a fine (taking away their money) for not obeying the road laws.

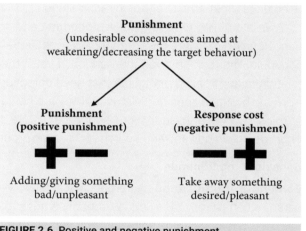

FIGURE 2.6 Positive and negative punishment

Ways to maximise the effectiveness of punishment

Effective punishment requires the following.

- It should be presented immediately following the undesirable behaviour to form a clear link between the behaviour and its consequence in the mind of the learner, especially in animals and young children.
- It should be the least painful punishment possible. Physical punishment may cause the recipient to feel fearful, resentful or aggressive towards their punisher. Response cost is a better form of punishment than the administration of physical or psychological pain.
- It should be perceived as aversive by the recipient and should be consistent in nature.
- It should be aimed at the behaviour rather than the person's personality.
- Once it has begun, the person applying it should *not* back down. Letting an individual out of a punishment negates the original message and negatively reinforces begging/pleading or other manipulative behaviours.
- It should be used in conjunction with positive reinforcement. Punishment alone may teach an individual to cease the behaviour that they are engaging in, but it fails to teach what the more desirable, appropriate behaviour should be. By also positively reinforcing the correct response, the individual is likely to behave more appropriately.

Elements of operant conditioning

As for classical conditioning, stimulus discrimination enables the individual to perceive the difference between a **discriminative stimulus (S^D)** and other stimuli, even if they are similar in nature. In operant conditioning, stimulus discrimination occurs when the person or animal makes the correct response to a stimulus for which reinforcement is obtained, but not to any other similar stimuli.

Stimulus generalisation, on the other hand, occurs when stimuli that are similar in nature to those for which reinforcement is obtained elicit the same response.

As stated earlier, extinction involves the termination of a CR. In operant conditioning, a CR will be extinguished if the reinforcement ceases to be given after it occurs. For this process to be done effectively, reinforcement must no longer be given at all, as partial reinforcement has been shown to increase and strengthen responses, thereby making the behaviour more resistant to extinction.

Spontaneous recovery occurs when a response reappears after extinction has occurred. Whether the behaviour continues depends on the consequence that follows it.

Applications of operant conditioning

Shaping in animal training

Shaping occurs when a reinforcer is issued after any response that is a successive approximation of a target response. This process ultimately results in this target behaviour being produced. Shaping is most effective in situations when the desired behaviour is unlikely to occur naturally. It has therefore been used most effectively in animal training.

Animal training typically involves operant conditioning techniques, especially the shaping of behaviour. Animals are rewarded for appropriate behaviours, initially on a continuous reinforcement schedule, and then on a variable schedule. Later, there is an emphasis on stimulus discrimination in relation to commands or objects that need detection. Animals are trained in a wide variety of contexts, from the home to zoos and circuses. They are not just taught to perform tricks but to fulfil a number of roles, such as guide dog work, support and care of paraplegics, police work (including the detection of drugs or explosives), and search and rescue procedures.

Behaviour modification

Behaviour modification systematically applies conditioning principles to adapt or modify a person's behaviour. Often used in school settings to eliminate inappropriate conduct in students, this methodology can be applied in any setting to help train people to develop skills or overcome undesirable habits or 'problem' behaviours. Within treatment contexts, such as mental health clinics or in psychological practice, the application of this procedure is referred to as 'behaviour therapy'.

A behavioural contract may be incorporated into such a process in which a written or verbal agreement between individuals stipulates desired behaviours, with rewards and punishments negotiated that are contingent on performing these behaviours.

Token economies

In a **token economy**, a tangible secondary reinforcement, such as a plastic chip, point or gold star, may be collected and exchanged for a primary reinforcement, such as food, or some other enticement, such as goods or privileges. These tokens are earned as a reward for demonstrating desired behaviour and may be withdrawn as a response cost or 'fine' for unsuitable actions.

Token economies can be seen in schools, prisons, psychiatric hospitals and 'sheltered workshops', often in conjunction with behaviour modification or behavioural contracts. Token economies also form the basis for a number of 'rewards programs' such as the 'shopper dockets' for fuel discounts, 'FlyBuys' and 'frequent flyers'.

2.1.4 Observational learning (modelling)

Observational learning, or **modelling**, is a social-cognitive approach to learning that occurs by watching others, noting the positive and negative consequences of their actions and/or then imitating these actions. Observational learning is of particular benefit when trying to learn or develop a motor skill. It is often easier and more efficient to imitate an observed *model* rather than try to learn through theory, which may not be totally clear and therefore misunderstood. It is also better than using 'trial-and-error' techniques, which may lead to ongoing mistakes and/or bad habits in the implementation of the task. Furthermore, learning through observation helps us to avoid potentially threatening and harmful situations without having to directly experience them.

Observational learning is often referred to as an extension of operant conditioning, as we learn through observing the consequences of others' behaviours, whether they are reinforced or punished, according to the processes outlined in operant conditioning. Because we learn from our interactions with others, Albert Bandura (1925–2021) also referred to his model as '**social learning theory**'.

FIGURE 2.7 The procedure in observational learning (as proposed by Bandura)

- **Attention**: In order for something to be learned, the individual needs to actively focus and concentrate while watching the model's behaviour. Furthermore, the behaviour should be clearly visible and perceived by the observer as being able to be imitated. A person is more likely to attend to models that are liked, known, similar in nature to the observer or have a high status.
- **Retention**: The observer must be able to remember the model's behaviour.
- **Reproduction**: The learner must attempt to reproduce or copy what has been observed in order to demonstrate that learning has occurred. This may depend on physical factors within an individual, such as their ability to perform the behaviour, and may also be influenced by environmental factors.
- **Motivation** and reinforcement: The learner has to want to perform the behaviour in order to receive some form of reinforcement. The behaviour should have a high incentive value for the learner or act as a reward, otherwise it will be unlikely that the behaviour will be carried out. Motivation may involve external reinforcement, self-reinforcement or vicarious reinforcement.

> **Note**
> *What does a pirate wear to protect himself?*
> To remember the order of the steps, think of the acronym '**ARRMR**' (pronounced as a pirate would say 'armour') – **A**ttention, **R**etention, **R**eproduction, **M**otivation and **R**einforcement.

- **Vicarious classical conditioning**: This occurs when an individual, having observed another person, copies their reaction to a specific stimulus.
- **Vicarious operant conditioning**: This refers to an individual learning by observing another individual being reinforced or punished for their behaviour. As a result, the individual will imitate this behaviour, behave in a modified manner or refrain from engaging in the behaviour, depending on the observed consequence for the model.
- **Vicarious reinforcement**: This occurs when the likelihood of an observer performing a specific behaviour increases after observing another being rewarded for such a behaviour.
- **Vicarious punishment**: This occurs when an observer is likely to refrain from performing a specific behaviour after observing another being punished for such a behaviour.

Bandura's research

Bandura formulated his social learning theory after performing a series of experiments with groups of kindergarten students (equal numbers of male and female) who watched films on a television screen. Having watched the films, the children were then individually placed in a room with a one-way mirror. The room contained several toys, including a BoBo doll (an inflatable rubber clown). The researchers then observed whether the children would shape their behaviour according to the aggressive model in the films.

In the first experiment, two groups of children observe an adult at play. For one group, the adult punches, kicks, hits and verbally abuses the BoBo doll, whereas for the second group the adult is non-aggressive in their play. A third (control) group does not observe an adult at play before being observed with the toys.

The main findings were as follows.

- Children who observed the aggressive model made significantly more aggressive responses than those in the other two groups.
- Boys made more aggressive responses than girls, especially if the model was male rather than female.
- The girls who observed the aggressive model also showed more physically aggressive responses if the model was male, but more verbally aggressive responses if the model was female.
- Not only were specific violent acts imitated, but the children also generalised their violence to other toys; for example, playing aggressively with plastic farm animals.
- The children who observed the non-aggressive model showed very little aggression, although not always significantly less than the control group.

Another experiment found that watching a film of aggressive behaviour inspired some imitation, but was less influential on children than observing the same aggressive act in person. This experiment was important, however, because it was a precedent that stimulated further studies into the effects on children of viewing violence in the media.

In a further experiment, three videos of an adult punching and abusing the BoBo doll were shown to different groups of kindergarten children. The videos differed in terms of the consequences given to the adult: rewarded, punished or no consequence. When allowed into the playroom, those who had seen the adult rewarded or receive no consequence were more likely to imitate the aggressive behaviour, whereas those who saw the adult punished showed significantly lower levels of aggression towards the doll.

Furthermore, the children who were then offered rewards for imitating the aggressive behaviour tended to behave aggressively, including those who had seen the adult model being punished for their actions. While the boys tended to show more overall aggression than the girls, when a reward was offered the girls became nearly as aggressive as the boys. On the basis of these results, Bandura concluded that all the kindergarten children learned about the aggressive model's behaviour, regardless of the consequences they observed the model face. They had made cognitive representations or mental pictures of what they had learned about behaving aggressively that could be elicited when an appropriate reward was offered. Therefore, the children acquired a learned response, which they stored until an incentive was offered.

Factors influencing imitation of behaviour

Bandura identified several factors as being important in determining whether an individual will copy an observed behaviour. These include the:

- amount of attention that the observer paid to the model
- characteristics of the model, such as similarity to the observer, attractiveness and trustworthiness
- observer's admiration for the model, in terms of status, expertise or power
- observer's ability to remember what the model did (retention)
- capabilities of the model
- observer's ability (or belief in their ability) to copy the modelled behaviour (reproduction)
- amount of motivation that the observer had to repeat a task
- consequence that the model incurred as a result of the behaviour – whether the model was reinforced or punished for the behaviour.

Applications of observational learning

Educational demonstrations

Observational learning techniques are often employed in instructional settings, especially in the development of specialised motor skills. Theory or verbal instruction may not be totally clear and therefore misunderstood, whereas watching an experienced model perform the actions is usually clearer and would prevent potentially threatening and harmful mistakes and/or bad habits in the implementation of the task.

This approach is employed in a variety of settings such as cooking classes, sports coaching, dance lessons and martial arts instruction.

Advertising

In addition to promoting products, advertising can be used to inform the public about contemporary social issues, such as mental health issues, alcohol and drug abuse, gambling addiction or domestic violence, with the aim of changing attitudes and promoting access to appropriate support services.

Other advertising is designed to inform the public about behaviours that should be avoided, such as the Transport Accident Commission's (TAC) campaigns dealing with a variety of dangerous driving behaviours that could lead to serious injury or death. These ads aim to prevent potentially threatening and harmful situations before drivers directly experience them, at which point it is too late.

2.1.5 Approaches that situate the learner within a system

Western learning styles tend to emphasise the role of a teacher as providing explicit instruction of abstract knowledge, which is often out of context in a classroom setting, followed by some form of objective assessment. In contrast to this, **situated learning** occurs within authentic settings and situations where key knowledge is integrated within an activity, context and culture.

For Aboriginal and Torres Strait Islander peoples, learning is situated on **Country**. The concept of Country means much more than a location. The term goes beyond a physical association with the land to incorporate the system of all living and non-living entities and ideas about law, spirituality, cultural practice and customs. Country is central to the identities of Aboriginal and Torres Strait Islander peoples as it represents the spiritual, physical, social and cultural interconnectedness between all creation. Knowledge can be described as 'being patterned on Country'. This captures the idea that knowledge is stored (situated) in the network of **songlines** that connect different places within Country.

In Aboriginal and Torres Strait Islander cultures, learning is embedded in relationships between people, defined through the system of kinship connections. Each person is situated within the kinship network and their position defines what knowledge they can hold and who they can share it with. Notice how this extends the standard Western ideas of what it means for learning to be situated.

Aboriginal and Torres Strait Islander approaches to learning are **multimodal**. This means that learning is communicated through and stored in many different formats. Storytelling, song, dance and art are all used to communicate knowledge. The stories are metaphors for important cultural contexts that make the information memorable. Knowledge is strengthened through expressions in song, dance and the use of cultural symbols in art. There is a strong emphasis placed on learning through observation and direct experience.

> **Note**
> Multimodal learning styles are also relevant in connection with Aboriginal and Torres Strait Islander memory techniques discussed in section 2.2.5.

2.1.6 Comparison of learning theories

TABLE 2.5 Comparison of relevance of antecedents and consequences within each learning theory

Antecedents (stimulus)	Classical conditioning		Observational learning
Behaviour (response)		Operant conditioning	
Consequences (result)			

TABLE 2.6 Comparison of learning theories

	Classical conditioning	**Operant conditioning**	**Observational learning**
Other names	Respondent conditioning Pavlovian conditioning Associative conditioning	Instrumental conditioning Skinnerian conditioning	Social learning Modelling Vicarious learning Vicarious conditioning
Key researcher(s)	Ivan Pavlov John B. Watson	B.F. Skinner Edward Thorndike	Albert Bandura
Basis of learning	Associating two stimuli such that either can trigger the same response	Associating a response with the consequence that follows it	Learning by closely watching someone else and then copying their behaviour
The source of the behaviour	Elicited by a stimulus	Emitted by the organism itself (while there may be some antecedent to the behaviour, the response is chosen by the organism according to the consequences of its actions)	Reproduced by the individual when sufficiently motivated to achieve the desired consequence
The type of behaviour (nature of the response)	Reflexive, involuntary behaviours. Responses are triggered by a specific stimulus or, after learning by association, a substitute stimulus. Responses could be physiological and/or emotional in nature. They could also lead to an aversive response to a particular stimulus and could develop into a phobia.	Predominantly non-reflexive, voluntary, intentional or spontaneous behaviours (although some involuntary, reflexive responses may be present). The behaviour occurs as a result of its consequences; that is, the subject is directed towards earning some sort of reward or avoiding/preventing some sort of penalty.	Voluntary, intentional behaviours. The behaviour occurs to receive some sort of reward or avoid/prevent an undesirable consequence.

TABLE 2.6 cont.

	Classical conditioning	Operant conditioning	Observational learning
The role of the learner	The learner is passive, as the response is caused by the UCS or, later, by the CS.	The learner is active, as the individual chooses to respond in a particular manner in order to lead to reinforcement or avoid punishment.	The learner is active, as the individual must pay attention to the model's behaviour, form a mental representation and decide whether or not the consequences are desirable.
The process of acquisition	After repeated pairings of two stimuli, an association is formed between them such that either stimulus can trigger the same response.	A voluntary behaviour is strengthened by association with a reinforcing consequence (a desirable outcome) leading to an increase in the response, or weakened by association with a punishing consequence.	The individual actively watches the model, paying close attention to their actions in order to retain this in their memory.
Stimulus discrimination	When an organism only exhibits the CR to a specific CS and not to other similar stimuli	When the organism only makes the correct response in a context that would lead to receiving the reinforcement, but not in relation to any other similar antecedents	When the individual only reproduces the behaviour in a context that would lead to receiving the reinforcement, but not at any other time
Stimulus generalisation	When a stimulus resembling the CS produces the same or a similar CR	When the organism makes the correct response following a stimulus that is similar to that for which reinforcement is obtained	When the individual reproduces the behaviour in a context that is similar to that for which reinforcement is obtained
The timing of the stimulus and response (timing of reinforcement)	Characterised by antecedent events. The stimulus (whether conditioned or unconditioned) always precedes the response (whether conditioned or unconditioned). The NS becomes a CS as it is presented just before (and thereby becomes associated with) the UCS. Reinforcement (the strengthening of the association) occurs before the response.	Reinforcement occurs after the response as a consequence of the behaviour. The antecedent (i.e. discriminant stimulus) occurs before the response, which then produces a consequence that will determine the nature of the learning and the likelihood of the response recurring.	Learning is not necessarily immediately evident. Reproduction of the observed behaviour could be delayed until desired/required. While behaviour is driven by the consequences, these may be perceived indirectly (vicariously) via observation rather than experienced directly (as in operant conditioning).
The effect of partial reinforcement	Likely to weaken the association between the UCS and the CS, thereby lessening the response rate to the CS	Can strengthen and increase the response rate as the organism pursues further reinforcement (as shown by the effect of schedules of reinforcement)	If reinforcement is not likely, the individual may not be sufficiently motivated to reproduce the behaviour
Extinction process	CR decreases when CS is repeatedly presented alone (i.e. without the UCS being present).	Responding decreases when the reinforcing consequences are removed.	Responding decreases if there is no reinforcement.

2.2 The psychobiological process of memory

2.2.1 Background knowledge underpinning models for explaining memory

Most researchers agree that **memory** involves an active information processing system that receives, organises, stores and recovers information.

While people emphasise retention of information as being the role of memory, there are in fact three key processes required: encoding, storage and retrieval.

Encoding, storage and retrieval

New information is encoded by converting it into a usable, meaningful form or code for input into the information processing system, thereby preparing it for integration and storage in long-term memory (LTM). Various methods for **encoding** information include making it meaningful, associating it with existing memories, using visual imagery and attaching personal references or emotions.

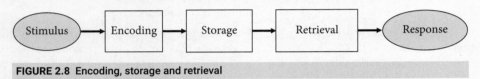

FIGURE 2.8 Encoding, storage and retrieval

Craik and Lockhart (1972) believed that it is the depth of processing that determines whether information is stored over a long rather than a short period. They defined depth in terms of a continuum based on the meaningfulness extracted from the stimulus rather than in terms of the number of analyses performed on it.

According to Craik and Lockhart, the three levels (types) of encoding (processing) are as follows.

1 *Structural (surface)*
   - Shallow processing on a superficial, perceptual level, focusing on the physical attributes or appearance of the information; for example, how the letters look or the typeface of a word
   - Items may be easily forgotten
2 *Acoustic (phonological, phonemic)*
   - Retention of stimulus according to how it sounds
   - Moderate level of processing
3 *Semantic (meaning)*
   - Attaching meaningful associations in order to understand the items to be remembered
   - Deepest level of processing

In addition to the depth of processing while encoding, Craik and Lockhart identified elaboration of processing and distinctiveness of processing as key determinants in the formation of LTM.

Storage entails the retention of information in memory over time. However, storage of information does not guarantee the individual's ability to retrieve it.

Retrieval involves recovering or accessing previously encoded information that has been stored in our LTM so that it can be used. This process depends on the manner of encoding and occurs best when retrieval cues match those present during encoding.

Information processing models of memory

Information processing models focus on people's ability to attend to their environment, encode information, store facts and retrieve knowledge. These theories assume that human information processing is comparable to computer processing, with input of information through the senses and processing occurring in stages that intervene between receiving a stimulus and producing a response.

This approach focuses on memory according to the type of processing, rather than the location of the information.

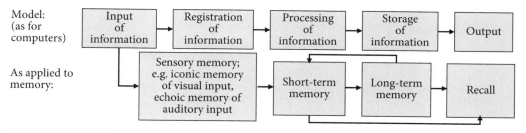

FIGURE 2.9 A cognitive (information processing) model of memory

2.2.2 Multi-store model of memory

At a time when computers were becoming popular, Atkinson and Shiffrin (1968) used computer programming as an analogy for their memory model, likening structural features to the computer and control processes to the person programming the computer.

Structural features are the fixed, inherent features of memory that do **not** change.
The basic structural features are:

- *three types of storage* – sensory, short term and long term
- **storage capacity** – the amount of information (number of items) that can be stored within the different levels of memory at any given time (*How much?*)
- **storage duration** – the length of time that information can be retained within the different levels of memory (*How long?*).

Control processes are the activities consciously chosen and utilised by people to assist in the memory process. The choice of control process depends on the situation.
Examples of control processes are:

- *attention* – selective focus on certain information
- *rehearsal* – moving information from short-term memory (STM) to LTM
- *retrieval* – which search strategy will be used to access the required stored information.

According to this information processing model of memory, there are three separate levels of memory that interact to enable encoding, storage and retrieval of information. These levels are sensory memory, short-term memory and long-term memory.

Sensory memory

Sensory memory is the process when items detected by the sensory receptors are retained temporarily in the sensory registers. These sensory registers have a large capacity for unprocessed information but are only able to momentarily preserve extremely accurate images ('traces') of sensory information – just long enough for relevant details to be attended to and transferred to STM. The function of sensory memory is to briefly save our sensory impressions so that a slight overlap occurs, thereby enabling us to perceive our environment in an uninterrupted fashion, rather than as a series of disjointed images and sounds.

The two types of sensory memory that have been most extensively explored are iconic (visual) memory and echoic (auditory) memory (Neisser, 1967).

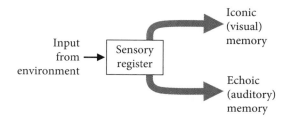

FIGURE 2.10 The two types of sensory memory

Iconic (visual) memory receives visual sensory information from the eyes and holds a brief visual image of what has just been perceived. The capacity is very large, limited only by the individual's field of vision. The duration of visual images in iconic memory is approximately 0.3 seconds, which allows 'smooth' perception rather than the blurring of moving objects.

> **Note**
> Iconic → begins with I → rhymes with 'eye' = visual
> This form of memory can be remembered by recalling that *icons* are *pictorial* in nature, either as graphics for computers or as artwork (especially religious paintings).

Echoic (auditory) memory is the sensory memory for audition (hearing); that is, for sounds that have just been perceived. The capacity is very large, limited only by the individual's range of hearing. The duration of sounds in echoic memory is approximately 3 to 4 seconds. A briefer interval would not allow enough time to attend to and process sensory information; for example, to link impressions of sound as speech to be able to make sense of the sounds as words or a sentence. This duration also acts as a filter because if the duration was longer, the words would start to overlap and be 'jumbled up' when we paid attention to them.

For example, someone talks to you while you are watching television and you respond to the sound of their voice by saying, 'What?' After a second, you suddenly realise what they said because you processed the echoic memory.

> **Note**
> Echoic → begins with E → just like Ear → **hearing/auditory**
> This form of memory can be remembered by recalling that **echoes** involve sound bouncing back from objects so that we can **hear** them again.

TABLE 2.7 Sensory memory

	Iconic (visual) memory	**Echoic (auditory) memory**
Capacity	Essentially that of the sensory system involved – very large (Sperling)	
Duration	Approximately 0.3 seconds	3–4 seconds
Processing	No additional processing beyond raw perceptual processing	

Attention

Loss of information from sensory memory can occur when one stimulus is quickly followed by another, which pushes the first out of sensory memory (Sperling, 1960). Information that is not quickly passed to STM is gone forever.

As an enormous amount of sensory information is constantly entering our sensory systems, we must focus our attention on a stimulus or mental event, thereby shutting out competing stimuli, in order to select what information will be processed within our STM.

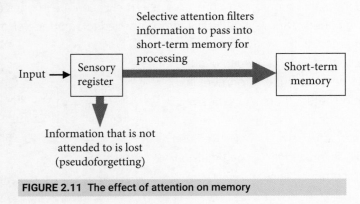

FIGURE 2.11 The effect of attention on memory

Short-term memory

Short-term memory (STM) allows an individual to consciously manipulate information contained in sensory or LTM. It has a limited capacity, and information is retained for a duration of approximately 18 to 20 seconds (although some sources say that information can occasionally last for up to 30 seconds). Because of this, STM is very susceptible to interruption or interference.

The *digit span test* is a measure of attention and STM that tests the recall of a series of digits. Findings from such tests show that STM has an average capacity of seven plus or minus two units of information (Miller, 1956).

> **Note**
> Short-Term memory holds Seven ± Two items.

Effect of chunking

An individual may increase the storage capacity of STM using **chunking**, the grouping of single units of information into higher-order units. This will allow an individual to remember between five and nine 'chunks' or slabs of information, as opposed to between five and nine individual units of information. For example, an individual may group the numbers 2, 0, 2, 3 into the year/number 2023. Chunking can also involve organising items into familiar, manageable units, such as the use of acronyms (e.g. BOLTSS = Borders, Orientation, Legend, Title, Scale, Source).

TABLE 2.8 Short-term memory

	Short-term memory (STM)
Capacity	Limited to approximately 7 ± 2 bits of information (Miller, 1956); this can be expanded by 'chunking' information into larger units
Duration	18–20 seconds (Peterson et al., 1959); some research findings suggest that information can occasionally last for 'up to 30 seconds'
Processing	To hold information in STM, it is often encoded verbally; however, other strategies such as visualisation may be used, making it possible to 'rehearse' the information

Rehearsal

Rehearsal maintains information in STM by preventing it from being lost or displaced by other material. The longer information is in STM, the greater the probability that it will be transferred into storage within LTM. The more frequently information is rehearsed, the better it is retained over time.

Maintenance rehearsal relies on the conscious recitation of information in a rote fashion. For example, repeating information over and over (usually subvocally in an individual's head) without adding any other meaning so that it can be kept in STM for longer than the usual maximum duration of approximately 20 seconds. Maintenance rehearsal is easily affected by distraction, which can displace information from our STM.

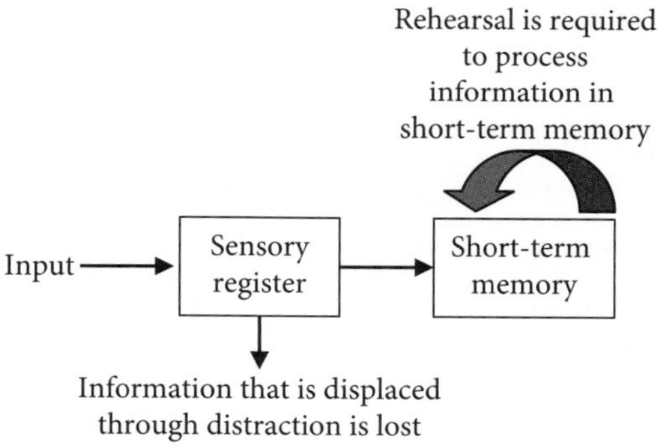

FIGURE 2.12 The effect of rehearsal on memory

In contrast, **elaborative rehearsal** involves the process of expanding on new information by adding to it or linking it with established material that an individual knows in their LTM. By adding further elements, elaboration makes this new information more meaningful, enabling deeper encoding that is more distinctive and unique, and therefore easier to locate during the retrieval process from LTM.

This may involve the analysis of semantic, sensory or physical attributes of the item to be remembered, which is then associated with items already stored in LTM or with other new information to aid encoding. The deeper the information is processed, the better it will be remembered.

This process forms the basis of most mnemonics (which will be discussed later in this chapter).

> **Note**
> When you are asked to *elaborate* about something, you need to *give/add* more information to make what you are saying more *meaningful*. The same is true with elaborative rehearsal.

Consolidation theory

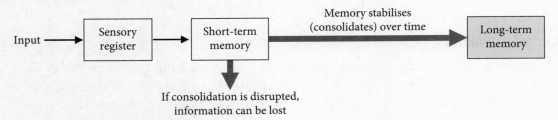

FIGURE 2.13 The effect of consolidation on memory

Consolidation theory proposes that, in order for new information to be transferred effectively from STM to LTM, there needs to be a time period during which these memories are able to fortify or stabilise without being disrupted. During this time, physical changes to the neurons occur to enable permanent storage of the new information. Consolidation takes at least 30 minutes on average. Once the memory is consolidated, it is relatively permanent.

If this consolidation period is disrupted, either by an accident or interference, the memory may be altered or completely lost.

> **Note**
> Con**solid**ation is when memories become **solid** (firm and fixed).

Long-term memory

Long-term memory (LTM) is a relatively permanent memory system that has an infinite capacity for storing information for an extensive period of time.

TABLE 2.9 Long-term memory

	Long-term memory
Capacity	Virtually unlimited
Duration	Up to a lifetime – relatively permanent
Processing	Information is organised according to meaning and is associatively linked

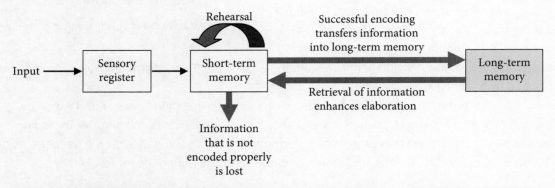

FIGURE 2.14 Processing information into long-term memory – encoding

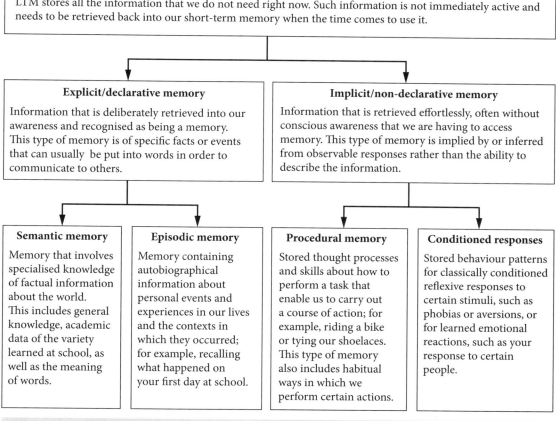

FIGURE 2.15 Types of long-term memory

Implicit and explicit memory

Explicit memory is memory *with* awareness. It contains information that can be consciously remembered, including episodic memory (life events) and semantic memory (words, ideas and concepts), and involves an intentional, deliberate attempt to retrieve previously stored information. Explicit memories are also called **declarative memories** because, if asked, you can consciously recall the information and can 'declare' or 'state' it. The most commonly used tests of explicit memory are recall and recognition.

Implicit memory, on the other hand, is memory *without* awareness and includes information that is unconsciously retained in memory, often learned without being aware that we have done so, but which still affects thoughts and behaviour. This type of memory is inferred from responses that can be observed rather than from the ability to describe the information stored here. As such, implicit memories are also referred to as **non-declarative memories** because people often find it difficult to 'declare' or 'state' the relevant information.

Implicit memory may include:

- habits
- automatic processes (such as responses learned through classical conditioning)
- **conditioned emotional responses**
- preferences and outlooks
- preverbal learning that takes place before a child or infant has developed a vocabulary
- motor skills.

Procedural memory is a function of implicit memory because we can perform a variety of tasks without consciously thinking about how to do so; for example, the ability to touch type would be stored in implicit memory, whereas reciting the letters on a particular row of the keyboard would require explicit memory.

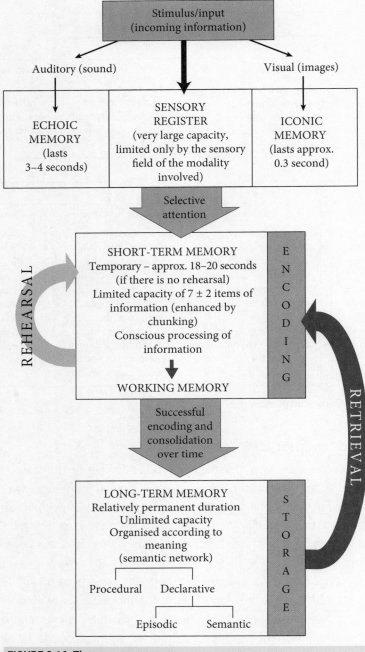

FIGURE 2.16 The memory process

2.2.3 Key brain regions involved in memory

No single, specific brain area handles all aspects of memory and, as with most cognitive activities, the cooperation of many brain areas is required for normal functioning.

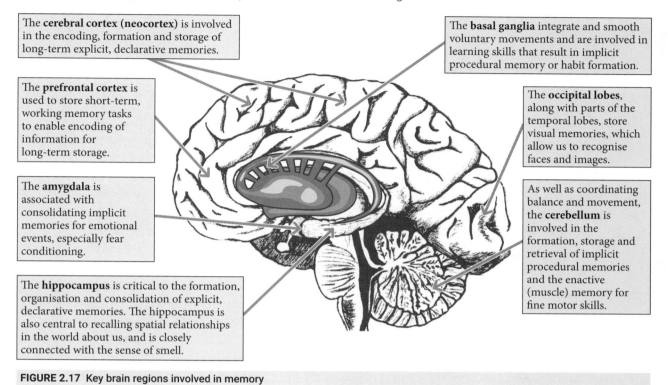

The **cerebral cortex (neocortex)** is involved in the encoding, formation and storage of long-term explicit, declarative memories.

The **prefrontal cortex** is used to store short-term, working memory tasks to enable encoding of information for long-term storage.

The **amygdala** is associated with consolidating implicit memories for emotional events, especially fear conditioning.

The **hippocampus** is critical to the formation, organisation and consolidation of explicit, declarative memories. The hippocampus is also central to recalling spatial relationships in the world about us, and is closely connected with the sense of smell.

The **basal ganglia** integrate and smooth voluntary movements and are involved in learning skills that result in implicit procedural memory or habit formation.

The **occipital lobes**, along with parts of the temporal lobes, store visual memories, which allow us to recognise faces and images.

As well as coordinating balance and movement, the **cerebellum** is involved in the formation, storage and retrieval of implicit procedural memories and the enactive (muscle) memory for fine motor skills.

FIGURE 2.17 Key brain regions involved in memory

Role of temporal lobes in memory formation

The key parts of the temporal lobes for memory are located in the **medial temporal lobes** (the inner part of this lobe towards the middle of the brain). They include the hippocampus, which plays a key role in the consolidation of conscious, declarative memories, and the amygdala, which is important for emotion-related, implicit memory.

Function of the hippocampus

The **hippocampus**, a structure within the interior of the brain wrapped around the thalamus and extending into the temporal lobes, plays a major role in memory, learning and recognition. Research suggests that the hippocampus is critical to the formation of explicit or declarative memories. It plays a part in deciding if information received by the senses is worth remembering, then mapping and organising memories before directing them to other sections of the **cerebral cortex** for storage, perhaps to several sections at once. The general role of the hippocampus in memory consolidation appears to be to provide a cross-referencing system that links and facilitates retrieval by drawing together all the different aspects of a memory from different areas of the brain.

The hippocampus is central to recalling spatial relationships in the world about us. Damage to it results in disorientation and an impaired ability to navigate our way in familiar settings.

The hippocampus is also closely connected with the sense of smell, helping to process signals from the olfactory bulb. This may account for the power of certain odours to evoke vivid memories and strong emotions.

Function of the amygdala

The **amygdala** is an almond-shaped set of neurons located deep in the brain's medial temporal lobes, adjacent to the hippocampus. The amygdala has a wide range of connections with other brain regions, allowing it to participate in a broad variety of behavioural functions.

In humans and other animals, the amygdala has been shown to play a key role in initiating and processing emotions such as fear responses, anger and pleasure. The hallmark of this memory system is that it is crucial for the acquisition and expression of fear conditioning, in which an NS acquires aversive properties as a result of being paired with an aversive event.

The amygdala is also involved in determining which memories are encoded and stored and where the memories are stored in the brain. It is thought that this determination is based on the level of emotional significance invoked by a stimulus or interpretation of events.

When a stressful situation is encountered, adrenaline is released into the bloodstream to act as a catalyst for the fight-or-flight response to react to the apparent stressor. Considerable evidence suggests that the sympathetic response also serves a long-term function by regulating memory storage for the emotionally arousing event, which then plays an adaptive role in the response of the organism to future situations.

It is well established that emotional experiences that induce the release of adrenal stress hormones also activate the amygdala, increasing its activity. This strengthens signals that result in the modulation of memory-related processes in other brain regions, such as the hippocampus. Considerable research evidence indicates that such activation can enhance the consolidation of declarative memories because of the association with positive or negative emotions. These studies reported that the relationship between activation of the amygdala during encoding and subsequent LTM was greatest for the most emotionally arousing stimuli.

Because the amygdala learns and stores information about emotional events, it is said to participate in emotional memory, which is viewed as an implicit or unconscious form of memory.

Function of basal ganglia and cerebellum

Procedural memories are encoded and stored via connections between the basal ganglia (especially the striatum) and the cerebellum.

The **basal ganglia** refers to a group of brain structures located beneath the cerebral cortex. These subcortical nuclei are responsible primarily for voluntary control of complex or skilled movement patterns, such as dancing or playing a musical instrument, as well as playing a key role in learning new motor skills involved in procedural memory and habit formation.

As well as coordinating balance and movement, the **cerebellum** is involved in the formation, storage and retrieval of implicit procedural memories and the enactive (muscle) memory for fine motor skills. Case studies involving anterograde amnesia have shown that such skills can still be developed and stored even if there is damage to the hippocampus, demonstrating that the cerebellum works independently in forming procedural memories.

2.2.4 Role of episodic and semantic memory

Semantic memory stores our knowledge of factual information, including general knowledge, academic information and the meaning of words. **Episodic memory** stores autobiographical information about personal events and experiences, including impressions of the context in which they occurred.

Although essentially different in nature, these two forms of LTM often work together when retrieving information, especially of autobiographical events or in imagining future possibilities. For example, when asked about your Year 12 formal, you would be able to say where and when it occurred (semantic) while describing events that happened from your viewpoint (episodic). Alternatively,, if asked to describe what you think your valedictory night will be like, you would call on your knowledge of how your school organises such events (semantic) and match these with images of similar events you have seen or attended in the past (episodic).

Not everybody, however, is able to call upon these faculties.

Alzheimer's disease

First described by Alois Alzheimer in 1906, **Alzheimer's disease** is an irreversible, progressive degeneration of nerve cells in the brain resulting in shrinkage of the brain tissue, particularly in the hippocampus and related areas of the midbrain, eventually leading to the death of the patient.

> **Note**
> Memory deficits are the first signs of Alzheimer's disease because of the deterioration of the *hippocampus*, which is responsible for consolidation of learning and recognition into memory.

Alzheimer's disease is the single most common cause of dementia, with deterioration of memory being the most widely recognised feature of the disease. The early stages of the disease are characterised by problems with STM, including forgetfulness and difficulty in forming new memories, as the hippocampus is affected first. Episodic memory is more noticeably affected, with an inability to remember autobiographical events, although there may also be problems with recent semantic memory.

As the disease spreads through the cerebral cortex, the patient gradually forgets how to do routine tasks and fails to recognise family members. Other features of the disorder include the patient becoming confused and disoriented about time and place, a decline in concentration and numerical abilities, an increase in frustration, unpredictable mood swings and emotional outbursts, changes in personality and language impairment. Progression of the disease leads to the death of more nerve cells and subsequent behavioural changes, such as wandering and agitation.

In the final stages, patients lose bowel and bladder control and need constant care. This stage of complete dependency may last for years before the patient dies. The average length of time from diagnosis to death is 4–8 years, although it can take 20 years or more for the disease to run its course.

Postmortem studies have identified the following effects of Alzheimer's disease on the physiology of the brain.

- Beta-amyloid protein plaques forming on axon terminals; that is, abnormal clusters of dead and dying nerve cells, the products of which have accumulated around sticky molecules of beta-amyloid protein. In Alzheimer's, these have been changed into a form that is toxic to the brain.
- Neurofibrillary tangles, which are insoluble, twisted structures caused by the build-up of protein and plaques within nerve cells that clog up the cell, affecting the brain's networks and its ability to retrieve information.
- Atrophy (wasting away) of brain tissue, especially deterioration of the hippocampus and eventual general brain shrinkage from cell death. At death, the brain of an Alzheimer's patient may have lost up to 50% of its weight.
- Destruction of neurons involved in the production of some neurotransmitters involved in memory, especially **acetylcholine**, which is necessary for cognitive functioning.

Currently, a definite diagnosis of Alzheimer's disease can only be made by examining the brain during a postmortem (after death) to confirm the presence of lesions due to the neurofibrillary tangles and abnormal plaque deposits. Brain scanning techniques such as magnetic resonance imaging (MRI), computerised tomography (CT) and positron emission tomography (PET) can reveal atrophy of the brain, and research is being conducted to assess their efficacy in identifying evidence of amyloid plaque build-up to facilitate earlier diagnosis of Alzheimer's disease. A working diagnosis can be made based on clinical observations and testing of cognitive capacity and memory loss, although deficits only become obvious once the condition has progressed and considerable neurological damage has already occurred.

The cause of Alzheimer's disease is unknown, but it probably results from a variety of genetic and environmental causes. While it is most common in the elderly, with the prevalence of Alzheimer's doubling every 5 years beyond age 65 and possibly affecting about one in four people over the age of 85, it is not a natural result of ageing, and early-onset cases have been reported for patients in their 30s and 40s.

There is currently no cure for Alzheimer's disease. The only available treatments are drugs that boost the efficiency of damaged neurons or ease some of the secondary symptoms of the disease, such as depression. New drugs are being developed that are aimed at inhibiting the production of toxic effects of the beta-amyloid protein plaques. Research evidence suggests that an active intellectual life, including regular social interaction, reduces the risk of Alzheimer's disease, or at least hinders its onset.

Aphantasia

Aphantasia is the inability to visualise images voluntarily in one's head. Typical symptoms include struggling to recall autobiographical memories as pictures, failure to recognise faces and difficulty in trying to visualise future events. It is estimated that 2–5% of people have the condition.

If you were to ask a person with aphantasia to imagine something, they could likely call on their semantic memory to describe the object, explain the concept and list facts that they know about the object. But they would not be able to experience any sort of mental image to accompany this knowledge.

The same thing would apply to autobiographical memories. While aphantasics can remember things from their past, they experience these memories in a different way than someone with strong imagery. If someone with aphantasia was asked to remember their last birthday, they would probably describe the event in semantic terms, such as who was there or the type of cake, but they will not be able to form a mental image to provide the episodic elements of the events of that day. Because of the effect on episodic memory, aphantasics could be less likely to be troubled by intrusive recollections or disturbing flashbacks where victims of trauma re-live these events in vivid sensory detail.

Independent studies using PET scans *and* functional MRI (fMRI) scanning have found that visual imaginings and things we *physically* see are processed in similar regions. Further to this, fMRI scans have shown that people with the ability to visualise vividly have a stronger connection between their visual network that processes what we see, and which becomes active during visual imagery, and the **prefrontal cortices**, involved in decision-making and attention. fMRI scans of individuals with aphantasia have found that when they tried to visualise imagery, there was a significant reduction in activation patterns across visual networks at the back of the brain, while frontal region activity responsible for decision-making and error prediction was significantly increased compared to controls.

Although aphantasia is a reduced visualising ability, those affected seem to have no trouble compensating for this with numerical or verbal replacements. Aphantasia is considered to be a variation of the human experience and not a condition that requires treatment.

2.2.5 Use of mnemonic devices

Cued recall aids in accessing information stored in memory by using a meaningful sign or hint to facilitate retrieval. A *retrieval cue* is a hint or suggestion that can initiate the retrieval process for accessing information from LTM. A cue will be an effective aid to retrieval if it is stored as part of the original memory.

A **mnemonic** device is a strategy or technique used to improve an individual's memory, usually using elaborative methods by adding information to the items to be remembered in order to make them more meaningful. Retrieval is simplified because organisation is enhanced. The information memorised is changed into a form in which it can link in or fit in more easily with the information already stored in LTM.

Some mnemonics employ *imagery* to enhance memory by linking a relevant mental picture, representation or idea with the concept to be remembered. For example, 'Stalac**tites** hold on **tight** to the roof of caves, whereas stalag**mites might** grow up to meet them', or 'pract**ice** contains **ice**, which is a noun, whereas pract**ise** contains **is**, which is a verb'. Another example would be for **echo**ic memory, as an **echo** is a sound or phrase that comes back in its literal form, but which fades quickly.

In each case, parts of the term are paired with images about the relevant concept or attribute to be remembered.

Association, on the other hand, refers to the assimilation of new material by relating or connecting it to already-learned information, thereby linking it into existing semantic networks in our LTM. The more things you can connect with a desired memory, the stronger that memory will be.

The repetitive use or rehearsal of memory-enhancing techniques helps us to recall the necessary information.

Acronyms

An **acronym** is a method of chunking information for retention by creating a pronounceable word using the first letters of a group of words; for example, WHO (World Health Organization), ANZAC (Australian and New Zealand Army Corps) and QANTAS (Queensland and Northern Territory Air Service). Examples within the course include SAME (sensory afferent motor efferent) for the types of neurons and BATheD (beta alpha theta delta) for the different brainwaves.

Some acronyms may form a series of words or appear to be a person's name, such as 'Roy G. Biv' (the colours in the visible spectrum). Through regular use, many acronyms have now become accepted as words themselves; for example, 'scuba' and 'laser'.

> **Note**
> An acronym must be pronounceable. The term is often misapplied to names of organisations, such as RACV or IBM. These are not acronyms, but initialisms, as they are said as a series of letters.

Acrostics

Acrostics involve the use of word associations where the first letter of each word in a list to be remembered is incorporated into a meaningful phrase or sentence that can be recalled more easily; for example, Some Actors Say Extra Rehearsals Store Long Renditions (the stages in the multi-store model of memory: sensory register, attention, STM, encoding, rehearsal, storage, LTM, retrieval). An acrostic is an effective tool for remembering the order of items in a sequence because elaborative rehearsal adds meaning and enhances encoding in declarative memory.

> **Note**
> Acro**nyms** use **i**nitials to form a **name** or word (the suffix '**-nym**' = word or name), whereas acro**stics** take **i**nitials from each **c**oncept to form a **s**entence.

Method of loci

Memory and spatial awareness are intertwined in the hippocampus, so people will remember things that happened at particular locations much better. This forms the basis of the mnemonic referred to as the **method of loci** (also known as the 'mind palace' or the 'memory palace').

> **Note**
> Loci → Location

This method uses a familiar sequence of places or locations (such as rooms in your home or landmarks on your way to school) or points on an easily imagined scene or figure (e.g. the back of your hand) as a series of cues that can be associated with the items to be recalled. This involves visualising an image linking each item to be remembered with each of the locations in the sequence. By retracing the route in your head and examining the images, you can cue recall of the items to reconstruct the concepts in order.

By linking items to each place in the sequences, you have absolutely grounded the list in order, so that you can't lose any of the items. This technique is best for serial recall of lists of items in a particular order.

> **Note**
> If you are allowed, apply this method by placing small posters of your notes around your house. Looking at these as you go about your daily routine not only revises the concepts to reinforce your memory, but also provides additional cues to help with retrieval.

Narrative chaining

Narrative chaining is a way of remembering a list of unrelated items by creating a story (or song) that links the concepts together in some ordered manner.

For example, a story to remember the words 'cat, ball, room, book, memory, essay, Macbeth, Friday' could be 'The cat played with a ball of wool in the living room, while her owner read a book about memory before writing an essay on Macbeth that was due on Friday'.

This mnemonic assists in memory by elaborating the information and enhancing its organisation in LTM.

> **Note**
> Narrative = story

Aboriginal and Torres Strait Islander memory techniques

Aboriginal and Torres Strait Islander peoples have the longest continuous cultural history of any group of people in the world. Their cultural knowledge has been preserved through powerful oral memory techniques that transmit knowledge through stories, song, dance and art. The knowledges held in these cultural practices are all deeply connected to places on Country. This is why connectedness to Country is so important to Aboriginal and Torres Strait Islander peoples.

In Aboriginal cultures, knowledge is stored in Country along Songlines. Songlines is an English word used to capture the Aboriginal cultural concept of ancestral knowledge being stored in sung narratives (stories) that are linked to sacred sites along routes used for trade and ceremonies. Songlines share some similarities with the method of loci, but they are not the same. Rather than being an isolated mnemonic technique, Songlines are integral to the cultural practices of Aboriginal peoples.

By singing the songs in the appropriate sequence, Aboriginal peoples could safely navigate long distances across land or sea. The sacred narratives and songs within the Songline connect Indigenous communities across large regions, forming a network that interweaves across the continent, and aiding travellers crossing the country by telling them where to find food, medicines, shelter or water by using significant topographical features. Aboriginal and Torres Strait Islander peoples also use the stars as memory spaces to represent certain geographical locations.

Some research has shown that the majority of the songs contain information to enhance survival, such as knowledge about the environment, land formations, seasonality, plant remedies and animal behaviour.

Songlines specific to a story continuum traverse the continent and intersect with others at specific locales, often indicated by an important landscape feature. Each location in the countryside acts as a mnemonic to a particular set of facts by associating certain information with that aspect of the environment, so the knowledge is literally grounded in the landscape. The process is supported by the use of portable devices, such as message sticks or different forms of art. Also, the types of imagery incorporated into the songs are often exaggerated and imaginative, as stories and mythologies are easier to remember than a list of facts. As one journeys along the route, at each key point or sacred site ceremonies are performed involving song, dance, story or ritual, all of which combine into a powerful memory technique that reinforces itself to help encode the information and further facilitate recall. In so doing, this technique goes beyond the method of loci as it incorporates other modalities, including narrative chaining, imagery and movement, to enhance the process. (Source: Margo Neale & Lynne Kelly, *Songlines: The Power and Promise*, Thames & Hudson, 2020.)

> **Note**
> While some suggested mnemonics have been given throughout this book, you should also come up with your own using any (or all) of the techniques mentioned. Creating your own mnemonics involves deeper processing which further enhances this process.

Glossary

acetylcholine A neurotransmitter involved in learning, memory, attention and motor control; found at significantly low levels in the brains of people suffering from Alzheimer's disease

acquisition The process of taking on a new behaviour, demonstrated by a consistent increase in responsiveness as a result of learning

acronym A method of chunking information for retention by creating a pronounceable word using the first letters of each word in a title or procedure (e.g. ANZAC)

acrostic A mneumonic device involving the creation of a memorable phrase in which the first letter of each word in the phrase matches the first letter of a term in a sequence of terms to be remembered, thereby acting as a retrieval cue for each of the required concepts

advertising The use of different forms of media to try to influence the attitudes and/or behaviour of a target audience

Alzheimer's disease A progressive and fatal neurodegenerative brain disease in which amyloid plaques and neurofibrillary tangles disrupt neural functions, causing cell death and atrophy of the brain. Early hippocampal damage disrupts consolidation of explicit memory first, followed by progressive loss of existing episodic and semantic memories due to neocortical damage

amygdala An almond-shaped brain structure located within each temporal lobe in front of the hippocampus; associates emotional information with explicit memories; plural amygdalae

antecedent An event or stimulus that precedes a particular response, indicating the likely consequence for the response and therefore influencing whether the response will occur

aphantasia A condition in which people suffer reduced or absent voluntary mental imagery

association The process of forming a connection between two or more stimuli, which can be developed through learning/conditioning

attention The active concentration of mental activity that involves narrowing the focus of awareness onto specific stimuli of interest or relevance while ignoring other stimuli. Within the formation of memory, this process determines which information will be processed further within short-term memory. In observational learning, a learner must watch a model closely in order to be able to remember their behaviour to replicate it later

A+ DIGITAL FLASHCARDS Revise this topic's key terms and concepts by scanning the QR code or typing the URL into your browser.

https://get.ga/aplus-vce-psych-u34

aversion therapy A form of behaviour therapy applying classical conditioning principles, whereby an individual learns to associate an undesirable behaviour pattern with an unpleasant response, so that they will experience this response in connection with the behaviour and therefore avoid doing it

avoidance conditioning A form of operant conditioning that occurs when an individual responds according to negative reinforcement in order to learn to evade or prevent an unpleasant event

basal ganglia A group of brain structures located at the base of the forebrain and in the midbrain that play important roles in controlling voluntary movement

behaviour The responses of an individual to externally or internally generated stimuli, including voluntary and involuntary responses

behaviour modification An application of the three-phase model of operant conditioning to help people change unwanted behaviours by modifying the antecedents that cue behaviours and the consequences that reinforce them

cerebellum The structure within the hindbrain beneath the rear of the cerebral cortex that coordinates balance and movement, along with the formation, storage and retrieval of implicit procedural memories and the enactive (muscle) memory for fine motor skills

cerebral cortex The layers of grey matter that cover the outside of the cerebral hemispheres, including multiple distinct functional regions associated with higher cognitive processes of attention, thought, perception, memory and language as well as sensory-motor processing; also called telencephalon

chunking An encoding mechanism that increases the capacity of short-term memory by grouping or combining multiple bits of information into a smaller number of meaningful units (e.g., ANZAC). This enables an individual to remember seven plus or minus two 'chunks' of information, rather than seven plus or minus two units of information; also strengthens storage and enhances retrieval from LTM

classical conditioning A type of learning in which an unconditioned stimulus that naturally evokes a specific unconditioned response is repeatedly paired with a neutral stimulus that does not usually produce this response. During this acquisition stage, an association forms between this neutral stimulus and the unconditioned stimulus. After conditioning, the once neutral (now conditioned) stimulus alone will elicit the conditioned response that it formerly did not produce

conditioned emotional responses A reaction that occurs when the autonomic nervous system responds to an emotionally provocative stimulus that did not previously elicit that response

conditioned response (CR) A reflex response to a conditioned stimulus in the absence of the unconditioned stimulus that would usually cause it

conditioned stimulus (CS) A previously neutral stimulus that acquires the ability to cause a reflex response through its association with an unconditioned stimulus

conditioning A term used to describe the process of learning when behaviours, events and stimuli become associated with each other

consequence In the three-phase model of operant conditioning, the feedback a learner receives from the environment as an outcome of a voluntary behaviour; can be reinforcing, punishing or neutral (no consequence)

contiguity Occurs in conditioning and refers to the association of two seemingly unrelated events when they occur close together in time and space

Country An Aboriginal and Torres Strait Islander concept of place as a system of interrelated living entities, including the learner, their family, communities and interrelationships with land, sky, waterways, geographical features, climate, animals and plants

declarative memory A subtype of long-term memory concerned with specific facts or events that can be brought consciously to mind and can usually be communicated to others. Declarative memory may be further divided into two subcategories: semantic memory and episodic memory

discriminative stimulus The stimulus (an event or object) that precedes a particular response, indicating the likely consequence for the response and therefore influencing whether the response will occur

echoic (auditory) memory A term used to describe auditory sensory memory traces; duration 3–4 seconds with very large capacity

elaborative rehearsal The association of new information with information that has already been stored in long-term memory or with other new information to aid encoding

encoding The processing of information in short-term memory to transfer it to long-term memory

episodic memory The component of explicit long-term memory used for storing and retrieving memories of personally experienced events and for imagining ourselves experiencing future events, accompanied by the feeling of mental time travel; also called episodic-autobiographical memory

escape conditioning A form of operant conditioning based on negative reinforcement, where a response that causes the cessation of an unpleasant stimulus increases in frequency over time

explicit memory (declarative memory) The kind of long-term memory we use when consciously remembering information about facts (semantic memory) or events (episodic memory)

extinction (of fear conditioning) The process of extinguishing (unlearning) a conditioned fear response through repeated presentation of the conditioned stimulus without the unconditioned stimulus

flooding A therapeutic technique, generally used to treat phobias, that exposes a patient to vast amounts of the feared conditioned stimulus in order to hasten the process of extinction

graduated exposure A therapeutic process whereby an individual is gradually taken, step by step, through a series of tasks to extinguish the conditioned response; also known as systematic desensitisation

hippocampus A seahorse-shaped brain structure within the temporal lobes of the brain associated with memory formation, encoding declarative information and transferring it from short-term memory to long-term memory to form and consolidate new declarative explicit memories

iconic (visual) memory A term used to describe the visual sensory memory system that registers visual information and holds it for approximately half a second.

implicit memory (non-declarative memory) The kind of long-term memory that is demonstrated through changes in behaviour and adaptive responses as a result of repetition or practice, without conscious recollection of the knowledge that underlies the performance; can operate independently of the hippocampus

law of effect Responses that result in desirable consequences are repeated, while those that result in undesirable consequences are not

learning The biological, cognitive and social processes through which an individual makes meaning from their experiences, resulting in long-lasting changes in their behaviour, skills and knowledge

long-term memory (LTM) The set of memory storage systems that enables us to store and retrieve knowledge and skills acquired over a lifetime with apparently unlimited capacity

maintenance rehearsal A short-term memory retention and encoding strategy that involves mentally repeating the names of items, ready for recall

medial temporal lobes The inner part of the temporal lobes towards the middle of the brain and includes the hippocampus, which plays a key role in the consolidation of conscious, declarative memories, and the amygdala, which is important for emotion-related, implicit memory

memory An active information-processing system that receives, organises, stores and recovers information

method of loci A mnemonic technique in which the items to be remembered are associated with specific locations on a familiar route or within a building, landscape or even the night sky

mnemonic Any device or technique used to assist encoding, storage and retrieval of memories, usually by creating an association between the information to be remembered and existing knowledge; also known as a memory aid

modelling An alternative term for observational learning, highlighting that the observer learns a new behaviour or modifies an existing one as a result of watching another person (the model) and copying their actions

motivation In observational learning, the cognitive processes that influence whether the learner decides to reproduce an observed behaviour based on their understanding of the observed consequences

multimodal A process that combines several different elements or approaches, such as multiple senses, kinaesthetic movement, art and so on

negative punishment The removal of a rewarding stimulus (reinforcer) as a consequence of a behaviour, making the behaviour less likely in future

negative reinforcement The removal of an aversive stimulus (punisher) as a consequence of a behaviour, making the behaviour more likely in future

neutral stimulus (NS) A stimulus (internal or external event) that does not naturally cause a reflex response

non-declarative memory This kind of long-term memory is often difficult to articulate, but is demonstrated through changes in behaviour and adaptive responses as a result of repetition or practice, without conscious recollection of the knowledge that underlies the performance; can operate independently of the hippocampus

observational learning A form of social learning in which the learner attends to the behaviours of another person, encodes the behaviours in memory and is motivated to rehearse and/or reproduce the behaviour based on their interpretation of the reinforcing consequences of the behaviour

operant conditioning A learning process in which the likelihood of a behaviour being repeated is determined by the consequences of that behaviour

positive punishment The addition of an aversive stimulus (punisher) as a consequence of a behaviour, making the behaviour less likely in future

positive reinforcement The addition of a rewarding stimulus (reinforcer) as a consequence of a behaviour, making the behaviour more likely in future

prefrontal cortex The most frontal region of the neocortex, involved in controlling attention and monitoring and regulating behaviour

procedural memory The kind of implicit long-term memory involved in learning and executing motor and cognitive skills that enable someone to carry out a course of action

punishment A consequence of behaviour that weakens the likelihood of the behaviour being reproduced

reflex An innate, automatic reaction, involving a simple, rapid response to a specific stimulus, which does not depend on conscious thought or processing by the brain

rehearsal The process of actively manipulating or repeating information in order to maintain it in short-term memory or aid its transfer into long-term memory

reinforcement A consequence of behaviour that strengthens the likelihood of the behaviour being reproduced.

reinforcer A rewarding stimulus that can be added or taken away as a consequence of behaviour, producing positive reinforcement and negative punishment, respectively

reproduction In observational learning, the cognitive process used to re-enact an observed behaviour or to rehearse it mentally

response A behaviour produced by an individual as an outcome of stimulus processing

response cost A form of punishment involving the removal of a desirable event or commodity that decreases the likelihood of a response recurring; also known as negative punishment

retention In observational learning, the cognitive process used to encode and store knowledge of observed behaviour

retrieval The process of bringing to mind knowledge of events or facts stored in explicit memory, or of initiating and executing an implicit procedural memory

semantic memory The component of explicit long-term memory that we use when we encode, store and retrieve factual and conceptual knowledge, and to recognise objects, people or places; accompanied by an awareness of knowing without a feeling of reliving the past

sensory memory The set of temporary memory stores with large capacity that enable sensory information to persist for a very brief duration so that goal-relevant information can be attended and encoded into short-term memory; includes iconic (visual) and echoic (auditory) sensory memory

shaping An operant conditioning procedure whereby a reinforcer is issued after any response that is a successive approximation of a target response. This process ultimately results in this target behaviour being produced

short-term memory (STM) A temporary memory store that represents information that is the current focus of attention, with limited capacity (7 ± 2 items) and a duration of several seconds, or for as long as information can be actively rehearsed

situated learning An approach to learning that occurs within authentic settings and situations where key knowledge is integrated within activity, context and culture, rather than via a teacher presenting abstract concepts out of context in a classroom setting

social learning theory An alternative term for observational learning that acknowledges the social-cognitive processes and the 'social context' in which they occur

Songlines The sung narratives encoded in physical routes across Country and in constellations in the night sky that convey Ancestral knowledge of Country; also known as song-spirals or Dreaming tracks

spontaneous recovery The sudden reappearance of a conditioned response after extinction had occurred

stimulus Any internal or external event that produces a response in an individual

stimulus discrimination The ability to perceive the difference between two or more stimuli, even if they are similar in nature. In *classical conditioning*, this occurs when an organism only exhibits the CR to a specific CS and not to other similar stimuli. In *operant conditioning*, this occurs when the organism only makes the correct response to a stimulus for which reinforcement is obtained, but not to any other similar stimuli

stimulus generalisation The likelihood that stimuli which are similar in nature will elicit the same response. In *classical conditioning*, this occurs when a stimulus resembling the conditioned stimulus produces the same or a similar conditioned response. In *operant conditioning*, this occurs when the organism makes the correct response to a stimulus which is similar to that for which reinforcement is obtained

storage The retention of information in memory over time

storage capacity How much information (number of items) can be stored within memory. The amount of data that can be held varies between the different levels of memory

storage duration How long information can be stored within memory. The length of time that data can be held varies between the different levels of memory

systematic desensitisation A behavioural therapy technique used in the treatment of disorders involving fear and anxiety, whereby an individual is gradually subjected, step by step, to increasingly similar stimuli to the conditioned stimulus itself through a series of tasks, to eventually extinguish the conditioned response

token economies An application of the three-phase model of operant conditioning where the learner receives tokens as reinforcers for desired behaviour that can then be exchanged later for actual rewards (e.g. FlyBuys points)

unconditioned response (UCR) An involuntary reflex response to a biologically significant stimulus

unconditioned stimulus (UCS) A biologically significant stimulus, such as food or a sudden loud sound, that causes a reflex response

vicarious classical conditioning Classical conditioning that occurs when an individual observes another's reaction to a specific stimulus

vicarious operant conditioning When a person learns by observing another individual being reinforced or punished for their behaviour. As a consequence, the observer will imitate this behaviour, behave in a modified manner or refrain from engaging in the behaviour

vicarious punishment Occurs when the likelihood of an observer performing a specific behaviour decreases after observing another being punished for such a behaviour

vicarious reinforcement Occurs when the likelihood of an observer performing a specific behaviour increases after observing another being rewarded for such a behaviour

Revision summary

Use the following summary of syllabus dot points and key knowledge within Unit 3 Area of Study 2 to ensure that you have reviewed the content thoroughly. Provide a brief definition or comment for each item to demonstrate your understanding or code them using the traffic light system: green (all good), amber (needs some review) or red (priority area to review). Alternatively, write a follow-up strategy.

How do people learn and remember?	
Models to explain learning	
• The concept of learning	
• Classical conditioning	
– Three-phase process	
– Neutral stimulus (NS)	
– Unconditioned stimulus (UCS)	
– Unconditioned response (UCR)	
– Conditioned stimulus (CS)	
– Conditioned response (CR)	
• Applications of classical conditioning	
• Operant conditioning	
– Three-phase process	

▪ Antecedents	
▪ Behaviour	
▪ Consequences	
– Positive reinforcement	
– Negative reinforcement	
– Positive punishment	
– Negative punishment (response cost)	
• Applications of operant conditioning	
• Observational learning	
– Social learning	
– Attention	
– Retention	
– Reproduction	

»	– Motivation	
	– Reinforcement	
	• Applications of observational learning	
	• Situated learning	
	– Aboriginal and Torres Strait Islander system patterned on Country	
	The psychobiological process of memory	
	• The concept of memory	
	– Encoding	
	– Storage	
	– Retrieval	
	• The multi-store model of memory (Atkinson–Shriffin)	
	– Sensory memory	
	– Short-term memory	
	– Long-term memory	
		»

• Rehearsal	
• Consolidation	
• Implicit memories	
• Explicit memories	
• The roles of specific regions of the brain involved in the storage of long-term memories	
– Cerebral cortex (neocortex)	
– Hippocampus	
– Amygdala	
– Basal ganglia	
– Cerebellum	
• The role of episodic and semantic memory in retrieving autobiographical events and constructing possible imagined futures	
• Alzheimer's disease	
– Brain imaging and postmortem studies of brain lesions	

• Aphantasia	
– Brain imaging studies of individual differences in the experience of mental imagery	
• The use of mnemonic devices	
– Acronyms	
– Acrostics	
– Method of loci	
– Aboriginal and Torres Strait Islander memory techniques	
– Songlines	

Exam practice

Models to explain learning

Answers start on page 229.

Multiple-choice questions

Question 1

Learning

A is any alteration in behaviour resulting from an event.

B may be temporary in nature.

C may be due to maturation.

D is a comparatively lasting change in behaviour because of past experience.

Question 2

Classical conditioning is said to have taken place when

A someone copies an action for which they saw someone else receiving approval.

B a stimulus automatically produces a specific response on the first exposure to it.

C a stimulus consistently produces a response even though it did not initially produce that response.

D a response increases in frequency as a result of its consequences.

Question 3

By pairing the ringing of a bell with the presentation of meat, Pavlov trained dogs to salivate to the sound of a bell, even when no meat was presented. Through his experiments, Pavlov demonstrated that the dogs developed an association between the

A neutral stimulus and the conditioned response.

B conditioned stimulus and the unconditioned stimulus.

C conditioned stimulus and the unconditioned response.

D neutral stimulus and the conditioned stimulus.

Question 4

In Pavlov's experiments with dogs, the food is referred to as the _____ and the salivation produced by the food as the _____.

A conditioned stimulus; conditioned response

B unconditioned response; unconditioned stimulus

C unconditioned stimulus; unconditioned response

D unconditioned stimulus; conditioned response

Question 5

In his experiments investigating the salivation response in dogs, Pavlov varied the timing between the sounding of a bell and the presentation of food.

Pavlov found that learning is quickest when the

A neutral stimulus occurs slightly before the unconditioned stimulus.

B unconditioned stimulus occurs slightly before the neutral stimulus.

C neutral stimulus is presented simultaneously with the unconditioned stimulus.

D unconditioned stimulus is produced after the neutral stimulus has been present for some time.

Question 6

According to Pavlov, an unconditioned stimulus is

A a behaviour acquired through the process of classical conditioning.

B an innate stimulus that has the capacity to elicit a reflexive response.

C an unlearned reaction to an environmental trigger.

D a previously neutral stimulus that, through association, has acquired the capacity to provoke a conditioned response.

Question 7

Loimata receives an injection for the first time and cries as a result of the pain. Several months later, she goes to the doctor again, and as soon as he gets out a syringe, Loimata cries. This is an example of

A a reflex.

B an operant response.

C spontaneous recovery of fear.

D a classically conditioned response.

Question 8

While preparing for his daughter's birthday party, Tahir keeps bursting the balloons as he is blowing them up too far. By the time he is inflating the fifth balloon, he notices that his wife is squinting and seems tense. It would appear that

A the noise from the bursting balloons has become a conditioned stimulus.

B Tahir's behaviour has become an unconditioned stimulus.

C inflation of the balloons to a large size has become a conditioned stimulus.

D his wife suffers from an anxiety disorder.

Question 9

After recovering from a serious motorcycle accident, Aldo is afraid to ride a motorcycle but not a bicycle. Aldo's behaviour best illustrates

A stimulus generalisation.

B secondary reinforcement.

C stimulus discrimination.

D extinction.

Question 10

In order to extinguish the conditioned response, Pavlov

A repeatedly presented the conditioned stimulus (bell) without the unconditioned stimulus (food).

B presented the unconditioned stimulus (food) without the conditioned stimulus (bell).

C paired the conditioned stimulus (bell) with the unconditioned stimulus (food).

D Either A or B.

Question 11

After being bitten by a stray dog as a child, Mila developed a dread of dogs. As some time had passed without having this reaction, she thought that she had gotten over her fear. Recently, however, while on her evening walk, she encountered a dog wandering on its own and experienced a sudden twinge of anxiety. This sudden anxiety best illustrates

A stimulus generalisation.

B spontaneous recovery.

C latent learning.

D extinction.

Question 12

Operant conditioning focuses on

A the way an individual makes associations between stimuli.

B how an individual copies the behavioural patterns of others.

C responses that are made to instructions given by significant others.

D the consequences that occur as a result of the behaviour.

Question 13

During _____ a pleasant stimulus should be given after the operant response has been made, whereas in _____ an aversive stimulus should be removed after the operant response has been made.

A positive reinforcement; negative punishment

B negative reinforcement; positive reinforcement

C positive reinforcement; positive punishment

D positive reinforcement; negative reinforcement

Question 14

Which one of the following is an example of negative reinforcement?

A Taking away a child's television privileges when they misbehave

B Punishing a child in front of another, to be an example

C Going to the dentist and having a toothache relieved

D Spanking a child for drawing on a newly painted wall

Question 15 ©VCAA 2014 SA Q46

Dan was born in England, but when he was five years old, his family moved to Australia. When Dan started school in Australia, the other boys teased him because they did not like his English accent; they wanted him to speak with an Australian accent. Dan quickly learned to speak with an Australian accent at school so that the boys would stop teasing him.

Dan quickly learned to speak with an Australian accent as a result of

A negative reinforcement.

B positive reinforcement.

C response cost.

D punishment.

Use the following information to answer Questions 16–18.

Jake is a chronic 'class clown' who constantly tells jokes, especially when the class laughs.

Question 16

As Jake's behaviour is now disrupting the learning environment for the other students, his teacher decides to punish him. The process of punishment

A requires the application of an undesirable consequence.

B increases the response rate.

C includes the withdrawal of an undesirable consequence.

D involves following the target behaviour with a beneficial result.

Question 17

Despite his teachers applying various strategies to discourage his disruption, from speaking angrily through to detentions, Jake's behaviour gets worse. It would appear that

A Jake has a severe learning disability.

B the punishment is perceived as a reinforcer.

C the punishment is having a generalised inhibiting effect.

D Jake's behaviour is getting worse as a way of getting back at his teachers.

Question 18

According to operant conditioning techniques, the best way to extinguish Jake's disruptive behaviour is

A the immediate application of an aversive punishment to the target behaviour.

B total cessation of any reinforcement for the behaviour.

C presentation of the conditioned stimulus without the unconditioned stimulus.

D for Jake to make a spontaneous, voluntary decision to discontinue the behaviour.

Question 19

Animal trainers teach dolphins to jump through a hoop by using a number of steps. First, they only give the dolphin some food whenever it approaches the hoop in the water. Then, after the dolphin does this regularly, they give the dolphin some food when it swims through the hoop in the water. Last, they hold the hoop progressively higher, rewarding the dolphin whenever it jumps through the hoop.

The process employed by the animal trainers to teach the dolphins applies the principles of

A trial-and-error learning.

B observational learning.

C graduated exposure.

D operant conditioning.

Question 20

Observational learning entails

A gaining knowledge through direct personal experience.

B listening to the verbal instructions of a more experienced individual.

C watching others.

D reflecting on the results of our actions.

Question 21

Which of the following accurately describes the order of the processes of observational learning outlined by Bandura?

A Motivation, attention, reproduction, retention, reinforcement

B Attention, retention, reproduction, motivation, reinforcement

C Motivation, attention, retention, reinforcement, reproduction

D Attention, motivation, retention, reinforcement, reproduction

Question 22

Children are shown a film featuring several aggressive acts by adults. Based on Bandura's research, you would expect them to

A be desensitised to violence in other movies.

B show no change in their subsequent patterns of play.

C engage in significantly more aggressive actions in subsequent play.

D engage in significantly fewer aggressive acts in subsequent play.

Question 23

After seeing the movie *Jaws*, depicting the panic caused by a great white shark ravaging swimmers at a coastal resort, many people were uneasy about, and even avoided, going swimming in the ocean for fear of being attacked. This response is best described as being due to

A a reflex action pattern.

B stimulus generalisation.

C higher order conditioning.

D vicarious classical conditioning.

Question 24

The various TAC advertisements that focus on different scenarios to promote better driving behaviour are relying on the process of _____ to be effective.

A vicarious operant conditioning

B continuous reinforcement

C classical conditioning

D insight learning

Short-answer questions

Question 25 (3 marks)

a What is the generally accepted definition of learning? 1 mark

b Explain the characteristics of a reflex action. Provide an appropriate example to support your answer. 2 marks

Question 26 (5 marks)

a Describe the types of responses that can be classically conditioned. 2 marks

b Outline how a behaviourist would describe a reflex using classical conditioning terminology. 1 mark

c Explain the characteristics of a conditioned stimulus. 2 marks

Question 27 (3 marks) ©VCAA 2004 E2 SB Q9

Aversion therapy was developed to deal with habits and addictions.

a Using the language of classical conditioning, describe an example of how this therapy may be used to help someone give up smoking. 2 marks

b Outline one major criticism of aversion therapy. 1 mark

Question 28 (4 marks)

a Explain 'operant conditioning'. 2 marks

b According to operant conditioning, how do different types of consequences affect our choice of behaviour? 2 marks

Question 29 (2 marks)

a Define 'reinforcement' within the context of operant conditioning. 1 mark

b Describe how the term 'punishment' is applied in operant conditioning. 1 mark

Question 30 (6 marks)

a The two types of reinforcement and punishment are distinguished as being either 'positive' or 'negative'. Clarify what the concepts 'positive' and 'negative' refer to within operant conditioning. 2 marks

b Explain, with the use of appropriate examples, the difference between negative reinforcement and punishment. 4 marks

Question 31 (4 marks)

Mike and Carol are new parents. When their baby cried at night, Mike would go into the baby's room, pick her up to comfort her and rock her until she went back to sleep. The baby continued to do this night after night and Mike got into the pattern of going to the baby's room as soon as she started to sob so that Carol could get some sleep.

Explain, using appropriate operant conditioning terminology, the pattern of behaviour for

a the baby. 2 marks

b Mike. 2 marks

Question 32 (8 marks) ©VCAA 2015 SB Q10

Every evening after school, Najida's father pestered her to do her homework. After a lot of pestering, Najida did her homework to make her father's pestering stop.

a Name the learning principle associated with Najida eventually choosing to complete her homework and state how this learning principle encouraged Najida to do her homework. 3 marks

b How could Najida's father use a token economy to encourage Najida to do her homework without pestering her? 3 marks

c Najida's mother decided she would withdraw Najida from her favourite weekend activity every time Najida did not do her homework.

Name the learning principle Najida's mother is applying and state why this may be effective in getting Najida to do her homework. 2 marks

Question 33 (5 marks) ©VCAA 2012 E2 SB Q2

Marco breeds birds and his favourite bird is Polly the parrot. Marco watched when Polly broke through her eggshell and hatched. Polly is now an adult bird. When Marco comes in to feed her, Polly gets very excited and screeches loudly when she hears the door of the birdcage open. Marco now wants to try to teach Polly to say some words.

a Identify one of Polly's behaviours in this scenario that is not due to learning. 1 mark

b Identify one of Polly's behaviours that indicates classical conditioning has occurred. 1 mark

c Identify one learning technique that Marco could use to teach Polly to say some words. Explain how Marco would use the elements of this technique to train Polly. 3 marks

Question 34 (10 marks)

a Explain why association is an essential element in the process of conditioning. Clarify how this process occurs within both forms of conditioning. 3 marks

b Compare the three learning theories with respect to the
   i type of behaviours involved. 2 marks
   ii role of the learner. 2 marks
   iii timing of the reinforcement. 3 marks

Question 35 (9 marks)

a Explain why Bandura's social learning theory is often referred to as a form of operant conditioning. 2 marks

b Describe two factors that could influence whether or not you would imitate a model's behaviour. 2 marks

c Seven-year-old Jyotika wants to make the tapestry that she received as a gift but has not done this sort of thing before. As she cannot understand the instructions, she asks her mother to show her how to sew long-stitch onto the background provided. Having watched her mother demonstrate the procedure a few times, Jyotika attempts to copy what she has seen. She shows what she has done to her mother, who praises her for doing a good job. Jyotika then continues with the rest of the tapestry on her own.

With reference to the scenario above, identify the way the steps of observational learning can be seen in practice. 5 marks

Question 36 (7 marks) ©VCAA 2020 SB Q4

Advertisers often use learning principles when promoting products. An advertisement for a new soft drink features people having a good time while consuming the product. This is intended to make potential customers experience positive emotions when thinking about the soft drink.

a What type of conditioning is used to generate positive emotions towards the new soft drink? Give two reasons to justify your response. 3 marks

b The next advertising campaign for the soft drink used a celebrity rather than an unknown person.

Identify two different processes involved in observational learning that demonstrate the advantage of using a celebrity to advertise the soft drink. Justify your response for each process. 4 marks

The psychobiological process of memory

Answers start on page 235.

Multiple-choice questions

Question 1
Which one of the following is **not** one of the three basic processes of memory?
A Attention
B Encoding
C Retention
D Retrieval

Question 2
Sensory memory
A is the initial stage of memory where external stimuli are registered by the sense organs and stored until processed into our long-term memory.
B is a temporary store of very limited capacity.
C saves information from our sensory receptors so that a slight overlap occurs, enabling us to perceive our environment in an uninterrupted fashion.
D is capable of storing visual and auditory information, but not touch.

Question 3
While you are listening to your teacher present a lesson in class, the process that determines what information moves from sensory memory to short-term memory is
A chunking.
B encoding.
C consolidation.
D selective attention.

Question 4
Short-term memory stores
A a limited amount of encoded information while it is required for further manipulation and processing.
B all forms of data, provided it is 'chunked' into larger units of information.
C only information retrieved from our long-term memory.
D all of the details gathered by our sensory systems until they are permanently encoded into our long-term memory.

Question 5
In order to demonstrate the capacity of short-term memory to his psychology class, a teacher applied a digit span test by reading out sequences of numbers of different lengths and getting his students to write them down after each set.

According to Miller's research, the average number of digits that can usually be retained within short-term memory is
A three.
B five.
C seven.
D nine.

Question 6

Participants in a research investigation are briefly shown a group of four random, unfamiliar words that they are to remember. They are then asked to count backwards by threes from 547. Which of the following would you expect?

A Recall of the words declines rapidly after about 4 seconds.

B Recall of the letters declines rapidly after about 20 seconds.

C Because 4 is less than the 'magic number 7 plus or minus 2', there is no forgetting.

D Since numbers and letters come from different categories, there is no retrograde interference and hence no forgetting.

Question 7

While recitation will help keep declarative information in short-term memory for longer periods of time, it does not necessarily facilitate transfer of that information into long-term memory. The most effective process to get more information into long-term memory is

A elaborative rehearsal.

B chunking.

C maintenance rehearsal.

D procedural encoding.

Question 8

Which type of memory retains autobiographical information and factual knowledge?

A Declarative memory

B Episodic memory

C Procedural memory

D Semantic memory

Question 9

Which of the following is **not** true of long-term memory?

A Its capacity is theoretically unlimited.

B It can hold information for an indefinite length of time.

C It includes both autobiographical and factual memories.

D It has about seven slots or chunks for information storage.

Question 10

Yuki still knows how to ride a bike even though she has not ridden one for several years. She can describe in vivid detail the steps she took in order to learn how to ride when she was a little girl, and recalls how she and her friends would have races along the street.

Yuki's retained ability to ride a bike would be stored in her

A declarative memory.

B episodic memory.

C procedural memory.

D semantic memory.

Question 11

'Consolidation' refers to the

A time necessary for stable short-term memory to form.

B period of time necessary for a lasting long-term memory to develop.

C process of forming relationships between objects or events for immersion into semantic networks within our long-term memory.

D process within which memory storage can be sped up through the application of electric shock.

Question 12

Identify the type of memory and the brain area involved in the storage of skills that enable us to carry out a course of action.

	Type of memory	Part of the brain
A	explicit	hippocampus
B	implicit	amygdala
C	episodic	amygdala
D	procedural	cerebellum

Question 13

The hippocampus is **not**

A closely connected with the sense of smell, helping to process signals from the olfactory bulb.

B central to recalling spatial relationships enabling us to navigate our way in familiar settings.

C clearly implicated in the formation and retrieval of implicit procedural memories.

D the provider of a cross-referencing system that links different aspects of a memory from around the brain.

Question 14

In the early stages of Alzheimer's disease, a common symptom is usually

A dizziness.

B forgetfulness.

C loss of appetite.

D loss of physical coordination.

Question 15

Which of the following statements about Alzheimer's disease is correct?

A Alzheimer's disease is a normal part of ageing.

B Scientists can pinpoint the exact cause of Alzheimer's disease.

C Alzheimer's disease is the most common form of dementia.

D When researchers examine the brains of people who have died of Alzheimer's, they find nothing unusual.

Question 16

A characteristic of the advanced stages of Alzheimer's disease is

A forgetfulness about how to do everyday things and an inability to store new long-term memories.

B significant interference with short-term memory function, but an ability to form new long-term memories.

C problems in storing new information in long-term memory, but continued access to old long-term memories.

D extreme difficulty in laying down new long-term memories, but no disruption to short-term memory.

Question 17

Aphantasia is the inability to

A form new episodic memories.

B describe semantic memories.

C use short-term memory to encode new long-term memories.

D visualise images voluntarily in one's head.

Question 18

Functional MRI scans of individuals with aphantasia found that when they tried to visualise imagery, there was

A a significant reduction in activation patterns compared to controls across cortical areas in the prefrontal regions responsible for decision-making, and in visual networks within the occipital lobes of the brain.

B a significant reduction in activation patterns in the prefrontal regions responsible for decision-making, while activity in visual networks within the occipital lobes of the brain was significantly increased compared to controls.

C a significant reduction in activation patterns across visual networks at the back of the brain, while frontal region activity responsible for decision-making and error prediction was significantly increased compared to controls.

D no significant difference in activation patterns across the brain compared to controls.

Question 19

The words SCUBA and LASER are examples of

A acrostics.

B acronyms.

C the method of loci.

D mnemonics that apply imagery to enhance memory.

Question 20

In preparation for an assessment task, two psychology students were trying to remember various terms within the sequence of memory formation. The students used different strategies. Alexi put up a series of posters along the path he took from his bedroom to the kitchen in his house and tried to memorise each as he went about his daily routine. Bethany made up a series of related sentences where each word began with the letters of the required concepts in their correct order.

The memory techniques used by the students were

A Alexi – acrostics; Bethany – Songlines.

B Alexi – acronyms; Bethany – acrostics.

C Alexi – method of loci; Bethany – acrostics.

D Alexi – method of loci; Bethany – acronyms.

Short-answer questions

Question 21 (7 marks)
a State the generally accepted definition of memory. — 1 mark
b Describe each of the three basic processes of memory. — 3 marks
c Explain why all three processes are required in memory. — 3 marks

Question 22 (9 marks)
List the three stages, in order, of Atkinson and Shiffrin's (1968) multi-store model of memory, describing the capacity and duration of each.

Question 23 (6 marks)
a What is the key function of sensory memory? — 1 mark
b Describe two attributes of iconic memory. — 2 marks
c Explain what is involved in echoic memory. — 2 marks
d Explain whether information can be rehearsed while in sensory memory to prolong its duration. — 1 mark

Question 24 (5 marks)
a Clarify where the information within our short-term memory comes from. — 2 marks
b Explain why it is beneficial for short-term memory to have a small storage capacity. — 2 marks
c Describe a method that could be employed to increase the capacity of your short-term memory. — 1 mark

Question 25 (9 marks)
a Explain how the process of rehearsal enhances the functioning of our short-term memory. — 3 marks
b Name and describe the form of rehearsal that has little, if any, effect on long-term memory formation. Provide one reason why this form of rehearsal is not effective for transferring information into long-term memory. — 3 marks
c Using an example, explain what is involved in the process of elaborative rehearsal. — 3 marks

Question 26 (6 marks)
Using appropriate terminology, distinguish between the various types of long-term memory.

Question 27 (5 marks)
Outline one of the roles that each of the following brain areas plays in memory, according to research.
a Hippocampus — 1 mark
b Amygdala — 1 mark
c Neocortex (cerebral cortex) — 1 mark
d Basal ganglia — 1 mark
e Cerebellum — 1 mark

Question 28 (4 marks)

a Describe the sort of information that is likely to be affected in the early stages of Alzheimer's disease. — 2 marks

b Identify two physiological effects of Alzheimer's disease on the brain. — 2 marks

Question 29 (4 marks)

a Explain how semantic and episodic memories work together in retrieving autobiographical memories. — 2 marks

b Outline how someone with aphantasia would go about describing one of their autobiographical memories. — 2 marks

Question 30 (3 marks) ©VCAA 2005 E2 SB Q5

Use an example to explain the mnemonic technique 'method of loci'. Describe the type of material that is most suitable for memorising using this technique.

Question 31 (5 marks) ©VCAA 2014 SB Q7

Laura must remember to stop at the following places in this order: post office, optometrist, supermarket and hotel.

a Give an example of an acrostic and an acronym that would assist her with remembering where she needs to stop. — 2 marks

b Explain how acrostics and acronyms assist in the retrieval of information from memory. — 3 marks

Question 32 (4 marks)

Explain how Aboriginal peoples' use of Songlines is indicative of the multimodal system of situated learning patterned on Country.

UNIT 4
HOW IS MENTAL WELLBEING SUPPORTED AND MAINTAINED?

Chapter 3
Area of Study 1: How does sleep affect mental
processes and behaviour? 100

Chapter 4
Area of Study 2: What influences mental wellbeing? 139

Chapter 5
Area of Study 3: How is scientific inquiry used
to investigate mental processes and
psychological functioning? 178

Chapter 3
Area of Study 1: How does sleep affect mental processes and behaviour?

Area of Study summary

Unit 4 applies a biopsychosocial approach to explore the influence of different factors affecting our physical and mental wellbeing.

This area of study investigates differing levels of consciousness, with a particular emphasis on sleep. As a naturally occurring circadian rhythm, biological mechanisms on the sleep–wake cycle are considered in relation to their influence on our physical and mental wellbeing. In connection to this, the effect of sleep deprivation on affective, behavioural and cognitive functioning is examined, including the findings of research comparing this state with different levels of blood alcohol concentration. Knowledge regarding zeitgebers will be applied to look at ways to improve the sleep–wake cycle.

Area of Study 1 Outcome 1

On completion of this outcome, you should be able to:

- analyse the demand for sleep
- evaluate the effects of sleep disruption on a person's psychological functioning.

The key science skills demonstrated in this outcome are:

- analyse and evaluate data and investigation methods
- construct evidence-based arguments and draw conclusions
- analyse, evaluate and communicate scientific ideas.

Adapted from *VCE Psychology Study Design (2023–2027)* © copyright 2022, Victorian Curriculum and Assessment Authority

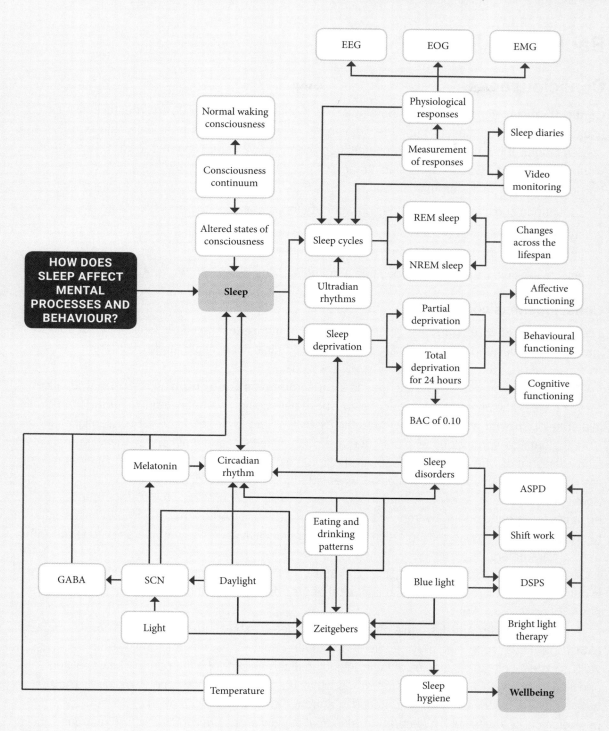

Background knowledge

Consciousness

Consciousness referrs to an individual's awareness of their own existence and mental activities (including thoughts, sensations and feelings) and of objects and events in the external world.

Sensations, thoughts, feelings and perceptions continually change and merge into one another without ceasing, like water in a flowing stream.

Consciousness is therefore:

1 continuous and changing, as it goes on without stopping and its contents blend into one another
2 subjective or personal, because it incorporates our perceptions of our internal feelings, thoughts and so on, and incorporates our immediate external environment, and
3 selective, as we can choose to focus on some things and ignore others.

Consciousness continuum

The term **state of consciousness** refers to the level of **awareness** that individuals have of both their internal processes and of external stimuli. Every individual experiences a range of different levels of consciousness throughout the progress of each day.

This progression is often represented as a **consciousness continuum**, where an individual can move between the different levels of consciousness depending on what is required at the time. While in a state of **normal waking consciousness (NWC)**, we are aware of the variety of sensory inputs from the outside world, along with our internal thoughts, feelings and reactions to these inputs. We are also cognisant of whether we are concentrating on the past, present or future. Levels of awareness may be high, as with *controlled processes* when attention is focused, or low, when routinely performing a well-practised task. Perceptions and thoughts are usually clear and logical. Bodily processes may be at the foreground of our awareness or operate automatically. Autonomic bodily processes, such as heart rate, respiration rate, galvanic skin response, brain activity and body temperature, operate within normal parameters.

As we spend the majority of our life in this state, most people assume without question that this is our experience of 'reality' and it is this state that provides the baseline by which we judge all other states of consciousness.

Any state that deviates from this baseline is considered to be an **altered state of consciousness (ASC)**. Cognitive processing may be affected, with alterations in the rate and quality of thought, as well as experiencing altered rules of logic, including disorganised thought, poor problem-solving and illogical thinking. *Perceptual distortions*, such as hallucinations, may occur. Individuals may have a disturbed sense of time, heightened or subdued emotions, or increased or decreased physiological changes such as heart rate, electrical activity of the brain, galvanic skin response and body temperature. Changes in self-control may also occur, such as feeling uninhibited and engaging in unusual or impossible behaviours. Additionally, an individual may experience a detachment from their own sense of identity. A person may experience an alteration in perceived body image and perceive bodily feelings not normally present, such as a change in the quality of energy flow in the body. Changes in continuity of memory over time may occur, with gaps in memory consistent with an altered state of consciousness.

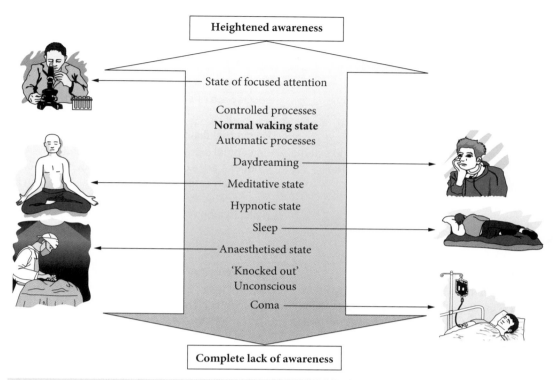

FIGURE 3.1 Levels within the continuum of consciousness

Naturally occurring states of consciousness are different levels of awareness along the consciousness continuum that normally occur as part of the natural daily cycle (e.g. daydreaming, sleep and dreaming). **Induced states of consciousness**, on the other hand, are different levels of awareness that have been purposefully generated or brought about by other factors (e.g. alcohol, medication/drugs) or due to external influences (e.g. meditation, hypnosis).

TABLE 3.1 Examples of naturally occurring and purposely induced states of consciousness

Naturally occurring states of consciousness	Induced states of consciousness
Daydreaming	Meditation
Sleep	Hypnotic state
Due to illness/fever	Anaesthetised
Connected with autonomic processes	Alcohol-affected state
Hyperstressed	Drug-induced state
	Highly focused attention
	Controlled processes

TABLE 3.2 Comparison of normal waking consciousness and altered states of consciousness

Normal waking consciousness	Altered states of consciousness and how they differ from normal waking consciousness
Awareness Aware of internal thoughts and feelings and of external stimuli through the senses.	**Awareness** Less aware of sensations and/or of external stimuli.
Attention Attention is focused on internal feelings and external stimuli and shifts of attention may be voluntary and involuntary. **Perception** Perception is clear, and the individual is able to process sensory input to form an awareness of their internal state and of external stimuli.	**Perceptual distortions** Perceptual distortions may occur. Senses/feelings/emotions are experienced as stronger and more vivid, or are suppressed and blurred. There can be a loss or detachment from a person's sense of 'self'. Hallucinations can occur.
Content limitations Content can be limited through selective attention to become organised and logical. Content remains for only a limited time as consciousness is a never-ending flow of sensations and perceptions.	**Content limitations** Content is not as limited, often becoming illogical, disorganised and nonsensical.
Cognitive processes Individuals are aware of their thoughts, which are clear and meaningful. Thoughts are structured, focused and flowing. Individuals are capable of analytical thinking.	**Cognitive distortions** Individuals may lose touch with reality. Information processing is also distorted. Thinking may be illogical and non-sequential, and difficulties in problem-solving and recall may be experienced.
Memory The brain actively stores information in memory and retrieves information from memory. Individuals are able to remember experiences and information processed in this state.	**Memory** Continuity in memory is disrupted, creating gaps or blackouts. It is often difficult to remember because information has not been processed into memory due to cognitive disruptions.
Time orientation Time is perceived as moving at a normal rate. Individuals also have an awareness of their place in time, and are able to focus on the past, present or future.	**Disturbed sense of time** Estimation of time is distorted. Time may appear to slow down, or pass more quickly (e.g. when meditating). Individuals may also have no sense of time (e.g. when asleep).
Emotional awareness Individuals are aware of their feelings and show a normal range of appropriate emotions.	**Changes in emotional feeling** Individuals become more emotional or emotionless. Inappropriate emotional reactions may occur and unpredictable emotional responses may result.
Levels of awareness There are various levels and degrees of awareness in normal waking consciousness. **Controlled processes** Information processing requiring conscious, alert awareness and mental effort in which the individual actively focuses their attention towards achieving a particular goal. **Automatic processes** Processes requiring little conscious mental effort and awareness, minimal attention and that don't interfere with other activities.	**Self-control** May experience difficulty in maintaining self-control, coordinating and controlling movements and maintaining control of emotions. Inhibitions are often lost, and individuals engage in unusual or risky behaviours. People become more open to suggestion than normal.
Autonomic processes Uncontrolled/autonomic processes operate within normal parameters.	**Autonomic processes** Autonomic processes increase or (more often) decrease in their level of activity.

3.1 The demand for sleep

3.1.1 Measurement of responses that can indicate different stages of sleep

The *self-report* method of measuring states of consciousness is a useful subjective measure for the researcher to gather data about the perceived experience, while the person is in that state, by asking the individual to describe it for them.

Sleep diaries

A **sleep diary** is a daily log that can be used to record an individual's sleep–wake pattern with related information, usually over a period of several weeks. It helps the individual to monitor when they go to bed and get up in the morning, how long it takes to fall asleep, waking patterns during the night and the quality of their sleep (how restful their sleep is). It also allows them to record any food, drink or activities that may be affecting their sleep. It is self-reported or can be recorded by a caregiver.

Benefits

- Thoughts and feelings of the individual are taken into account
- Gives the perspective of what the individual experiences
- Results from video monitoring can easily be used in conjunction with other results
- Time-effective

Limitations

- Results are highly subjective – difficult to compare across people
- Results depend on the reliability of the individual – requires conscious awareness and an ability to describe their experiences in words
- Hard to remember all details/accuracy in the morning
- May neglect, intentionally or unintentionally, to provide crucial information to the researcher
- Results are open to individual interpretation or bias
- Participant may be dishonest

Behavioural observations

Behavioural observations are useful in demonstrating some of the effects of altered states, but they are limited because they tell the researcher very little about what is actually happening inside the body.

Video monitoring

Video monitoring uses video cameras to record externally observable (or audible) physiological responses of the individual during the night.

Very little or no body movement indicates REM (rapid eye movement) sleep. An increase in movement indicates NREM (non-REM) sleep. As the cameras are silent, they should not disturb the individual's sleeping patterns. Video monitoring can also be done in the patient's home, allowing a more natural environment.

Benefits

- The individual is unlikely to be disturbed
- They can be monitored at home or in a lab
- Good for observations of actual behaviour

- Results from video monitoring can easily be used in conjunction with other results, such as physiological measurement
- Due to the nature of video monitoring, results can be kept for use at a later date

Limitations

- Results may be limited/inconclusive
- Results may be subjective and open to different interpretations
- Finding results may be time-consuming

Physiological responses

Physiological responses give a better indication of differing states of consciousness because they are objective measures that also provide information about bodily functions in those states.

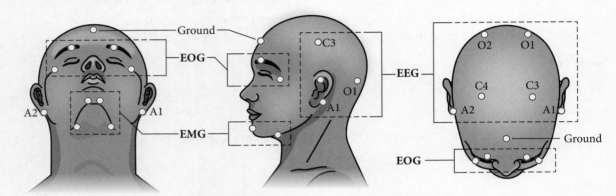

FIGURE 3.2 Locations for placement of electrodes for EEG, EMG and EOG when used in sleep research

Muscle tension

The **electromyograph (EMG)** detects, amplifies and records electrical activity in muscles, generally via electrodes attached to the jaw or lower face. It is used in sleep laboratories to measure muscle tension, which is extremely lax (to the point of paralysis) during REM sleep.

Lowered levels of muscle tension can also be evident in other altered states, such as relaxation or meditation.

Eye movement

An **electro-oculograph (EOG)** measures eye movements by detecting, amplifying and recording electrical activity in the muscles around the eyes via electrodes attached to the upper face, usually one above each eye and one just to the side of each eye near the temple. Rapid eye movements indicate that an individual is in REM sleep.

Electrical activity of the brain

Invented by Berger (1929), the **electroencephalograph (EEG)** is a device that measures electrical activity in localised areas of the brain. Because neurons use electricity to communicate, the EEG records patterns of electrical impulses produced by neurons activated in the brain. These impulses are measured via a series of electrodes placed at various points on the surface of the subject's scalp. The changes in electrical activity of hundreds of thousands

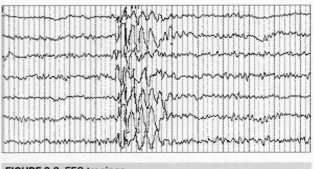

FIGURE 3.3 EEG tracings

of neurons within the brain in the vicinity of each electrode are detected, amplified and transferred to paper to be recorded as brain wave tracings known as an *electroencephalogram*.

Changes in electrical activity are particularly evident in behavioural states such as sleep, wakefulness and arousal, and abnormal electrical activity can signal disease states such as epilepsy and coma.

Brain waves can differ from normal waking consciousness to altered states. The normal waking state is characterised by high-frequency, low-voltage **beta-wave** patterns. Slightly larger **alpha waves** are present when we are daydreaming, in a relaxed or meditative state, or as we are drifting off to sleep. Other progressively larger brain waves (**theta waves** and **delta waves**) can be observed during sleep, which is why EEGs are used in sleep laboratories.

> **Note**
> To remember the brain waves as we descend from being awake to deep sleep, think of the acronym: **BATheD** (**B**eta, **A**lpha, **T**heta, **D**elta).

Advantages of EEG

- Non-invasive technique
- General measure of brain activity
- Can differentiate between different neurological conditions or behavioural states
- Inexpensive and safe
- Can be used for research with healthy and clinical participants

Limitations of EEG

- Time-consuming technique
- General measure of activity:
    - averaging the activity from millions of neurons cannot establish precise activity in a particular, localised region
    - signals measured are recorded on the scalp, which means it only maps the surface of the cortex and may not represent the activity in the underlying cortex

> **Note**
> The methods used to measure physiological responses in sleep begin with **E** (electro), which refer to the measurement of electrical impulses, and finish with **G** (graph), referring to representation of the impulses in graphical form.
>
> The central part of each term relates to the specific physiological response being measured by the device.
>
> EEG: **E**ncephala = Brain (you could also make an association with encephalitis, inflammation of the brain, or hydrocephalus, water on the brain)
>
> EMG: **M**yo = Muscle
>
> EOG: **O**cular = Eyes (you might remember this term from Unit 2, or the word 'optic', which also begins with the same letter, may reinforce the association with eyes).

> **Note**
> When defining each method within a short-answer question, remember **DARE**: each device **D**etects, **A**mplifies and **R**ecords **E**lectrical activity.

Identifying EEG brain wave patterns

Amplitude refers to the height or size of the peaks and troughs, indicating the level of voltage within the electrical impulses.

Frequency refers to the number of brain waves per second.

TABLE 3.3 Brain waves associated with different states of consciousness

Brain wave	Frequency	Description	State of consciousness
Beta	High (12 to 40 Hz)	Fast (high-frequency, low-amplitude) waves normally produced when your brain is in an active state of awareness, such as when you are actively engaging in a thought process or trying to solve a problem.	Awake Alert Stressed Engaged in active thought Focused attention REM sleep
Alpha	Relatively high (8 to 12 Hz)	Brain waves (slightly bigger than beta waves) that are produced during relaxation, mostly with eyes closed and upon deep self-introspection when your brain is in a 'quiet awareness' state of mind. Your brain produces more alpha waves with your eyes closed because you are not distracted by the outside world and you are better able to focus internally.	Relaxed Daydreaming Light meditation Drowsiness Hypnogogic state NREM stage 1 Hypnosis
Theta	Medium (4 to 8 Hz)	Medium-sized brain waves usually associated with sleep, but also possible in a highly internalised state of consciousness.	Starting to fall asleep / light sleep NREM stages 1–3 Deep meditation
Delta	Low (1 to 4 Hz)	The slowest (low-frequency, high-amplitude) brain wave associated with deep, restorative sleep.	Deep sleep NREM stage 3 Anaesthetised

TABLE 3.4 Physiological measures associated with different states of consciousness

Sleep stage	EEG pattern	EMG pattern	EOG pattern
Awake (NWC)	An awake, alert state with beta waves (high frequency, low amplitude).	High amplitude, often with distortions	Active; may be spiky with eye blinks
NREM stage 1 (drowsy)	There is a decrease in alpha waves from the awake, resting state, and more medium-frequency, irregular amplitude theta waves are present.	High amplitude, but without distortions	May be rolling eye movements
NREM stage 2 (light sleep)	Theta waves are lower in frequency and higher in amplitude than they would be in stage 1. Sleep spindles (high-frequency bursts) and K complexes are also present. Most researchers say 'true sleep' has not begun until after the first spindle occurs.	Medium amplitude	No eye movements
NREM stage 3 (deep sleep)	Delta waves are very low in frequency and very high in amplitude.	Medium or low amplitude	No eye movements
REM (rapid eye movement)	Irregular bursts of high-frequency, low-amplitude beta-like waves.	Low amplitude	Sharp, intermittent eye movements

> **Note**
>
> To recall the size of the waves as we descend through them, imagine that you are in a highly active/agitated state and have to draw brain waves. Without lifting your pen, gradually allow yourself to relax.
>
> The shape of your waves would be similar to the diagram on the right.

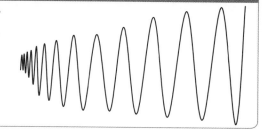

3.1.2 Sleep

Sleep is an altered state of consciousness marked by reduced metabolism and lowered consciousness. Sleep consists of different stages, each of which can be distinguished by characteristic physiological responses, including brain wave patterns, the presence or absence of rapid eye movement, and changes in heart rate, breathing rate, body temperature and muscle tone.

NREM (non-REM) sleep incorporates those stages of sleep not associated with rapid eye movements, and contain little dreaming. During NREM sleep, an individual is able to move, body temperature falls, breathing and heartbeat are regular, and brain waves are slow and rhythmical. We have little conscious experience at these times.

NREM sleep consists of several stages, designated by changes in brain wave patterns, ranging from drowsiness through to deep sleep. In the early stages (stages 1 and 2), you awaken easily and may not even realise that you have been sleeping. In deeper sleep (stage 3), it is very difficult to wake up and if you are aroused, you are likely to find yourself disoriented and confused. In NREM sleep, your muscles are more relaxed than when you are awake, but you are able to move (although you tend not to because the brain is not sending signals to the muscles to move).

REM (rapid eye movement) sleep, on the other hand, begins about 90 minutes after we fall asleep and is distinguished by rapid eye movements and inhibition of voluntary muscles despite electrical brain activity being similar to that observed during the waking state.

Key characteristics of the different stages within the **sleep–wake cycle** are summarised in Table 3.5.

TABLE 3.5 Characteristics of stages within the sleep–wake cycle

Stage of the sleep–wake cycle	Description
Awake	A state of normal waking consciousness indicated by awareness and an ability to process our thoughts clearly. Our eyes are open and we can interact with the external world. *Beta waves* are recorded by an EEG.
Hypnagogic state	The dreamlike 'twilight' state at the beginning of our night's sleep between wakefulness and entering NREM stage 1, which lasts about 1–2 minutes. During this state: • we begin to lose voluntary control over body movements • our sensitivity to external stimuli diminishes • our thoughts become more disorganised, fanciful and less bound by reality.

TABLE 3.5 cont.

Stage of the sleep–wake cycle	Description
NREM stage 1	A period of 'light sleep' as we transition from wakefulness into true sleep. During NREM stage 1: • breathing is irregular • heart rate decreases • eye movements slow • as muscles relax, **hypnic jerks** can occur • brain waves are irregular and small, with *alpha brain waves* getting progressively bigger to form medium-sized *theta waves*. A person can be easily aroused by external stimuli. If woken, the person may think they haven't been asleep.
NREM stage 2	The stage considered to be the transition into 'true' sleep where people are unlikely to respond to stimuli. While more difficult to waken, the individual is still relatively easy to arouse, at which point they are likely to say that they were just 'dozing'. During NREM stage 2: • body temperature is lowered • eye movements stop completely • breathing and heart rate slow further and become more regular • muscles become more relaxed • you become less aware of your surroundings. The following distinguishing features also occur within the *theta brain waves*: • **sleep spindles** (brief high-frequency bursts of brain activity important for learning and memory), and • occasional **K complexes** (high-amplitude spikes that occur in response to environmental stimuli).
NREM stage 3	Considered to be 'deep sleep', this stage is also referred to as 'slow wave sleep' (SWS) as theta waves transition to become the very slow, large amplitude *delta waves*. During NREM stage 3: • muscles are very relaxed • having dropped to their lowest levels, heartbeat/blood pressure and breathing rates are slow and regular • human growth hormone is released to enable the physical growth and repair of the body and muscles after the stresses of the day, and may help recuperation from physical activity or illness • reduced brain activity during SWS allows the glymphatic system to use cerebrospinal fluid to flush out accumulated toxic waste products from our brain • an individual is likely to be unresponsive to any environmental stimuli and is very difficult to wake up. If woken during this stage, individuals usually experience **sleep inertia**, where they feel 'groggy', confused and disoriented and have a poor memory of sleep events. Periods of stage 3 occur earlier in the night's sleep.

TABLE 3.5 cont.

Stage of the sleep–wake cycle	Description
REM	REM sleep begins about 90 minutes after we fall asleep and returns four or five times a night, becoming longer and more frequent throughout the night. Individuals are difficult to waken from REM. Characteristics of REM include the following: • bursts of rapid eye movements back and forth under your eyelids • faster, irregular, shallow breathing • an increased heart rate • significant brain activity, as shown by erratic high-frequency, low-voltage beta-like brain wave patterns (similar to those observed during the waking state) • limp and immobilised muscles due to nerve impulses to body muscles below the neck being blocked. REM sleep is sometimes referred to as **paradoxical sleep** because the brain is showing a high level of activity while the body is inactive and extremely relaxed (to the point of temporary paralysis). Studies suggest that REM sleep may help enhance brain development, facilitating the formation of synapses and strengthening neural connections in the brain to enable the consolidation of long-term memories. Most vivid night dreams occur during REM.
Hypnopompic state	The final stage of the sleep cycle leading out of REM sleep as the individual wakes up.

> **Note**
> Previous Study Designs described NREM stage 3 as two separate stages (NREM 3 and 4). Be aware of this when looking at past exam questions and hypnograms.

Pattern of the stages of sleep

An adult sleeps for 7–8 hours per night, progressing through ultradian **sleep cycles** of about 90 minutes.

Adults spend approximately 80% of their night's sleep in NREM and, typically, the first half of the night has more NREM sleep than the second half. Periods of **deep sleep** (NREM3) occur earlier in the night, and the depth of NREM sleep gets progressively shallower with each cycle as the night goes on.

Periods of REM sleep occur, on average, every 90 minutes. An adult experiences a total of 1–2 hours REM sleep a night, in four to five sessions, with each session progressively increasing in length as the night goes on, starting from 10 minutes in length to 30 or more minutes in the morning.

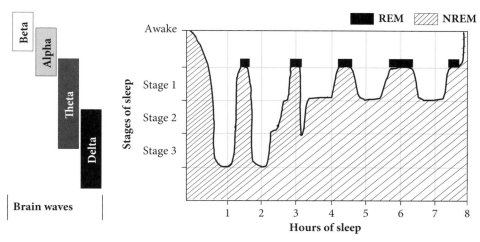

FIGURE 3.4 The sleep cycle of a young adult

> **Hint**
> To recall the cycles of sleep, remember the following.
> 1. There is an old adage that says that every hour of sleep before midnight is worth double those after midnight.
>   → Our deepest sleep occurs early in the sleep cycle, not later.
> 2. We remember more of our dreams that occur just before we wake up after a night's sleep.
>   → Dreams get longer through the night and occur in REM sleep as we cycle through NREM stage 1 (light sleep).

3.1.3 The biological mechanisms of sleep

Biological rhythms are internal mechanisms that synchronise physiological activity.

Circadian rhythms are biological cycles that occur about every 24 hours. (*Circa diem* is Latin for about a day.) Examples would be our daily sleep–wake cycle, body temperature and certain hormones, including cortisol and melatonin. These rhythms combine to form what is often referred to as our 'body clock'.

Ultradian rhythms are biological cycles that occur more than once a day. An example would be the NREM/REM cycles of about 90 minutes within the sleep period.

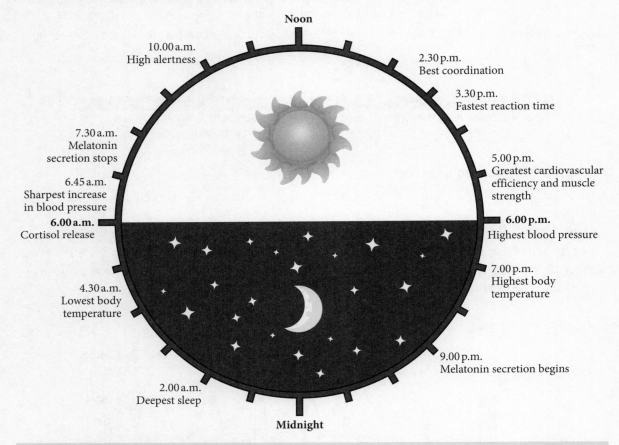

FIGURE 3.5 The circadian body clock

The circadian clock affects the daily rhythms of many physiological processes. **Zeitgebers** (from the German for 'time giver') are environmental cues that regulate and **entrain** (determine or modify the phase or period of) circadian cycles. These external cues synchronise the internal body clock with the world around us.

In humans, **light** is the main environmental cue that influences the sleep–wake cycle. The daily schedule of sleeping and waking is regulated by the **suprachiasmatic nucleus (SCN)** within the brain's hypothalamus. In the evening, when light levels decrease, the SCN sends signals to the **pineal gland** to release **melatonin**, a neurohormone that initiates sleepiness when its levels increase. In the morning (after sunrise), when light stimulates receptors in the retina of the eye, the receptors send signals to the SCN, triggering decreased output of sleep-inducing melatonin by the pineal gland and increased levels of cortisol to help us wake up.

Although circadian rhythms tend to be synchronised with cycles of light and dark through the regulation of the SCN, other factors can influence the timing as well, such as ambient temperature, eating or drinking patterns, work or school schedules, exercise and additional lifestyle choices.

Temperature changes may act as a zeitgeber. Fluctuations in ambient temperature occur in connection with the day–night cycle; warmer in the day when we should be active and alert, and colder at night when we should be sleeping. However, because of daily and seasonal changes, temperature cues are less obvious.

For humans, activities defined by timetables or socially defined routines, such as regular mealtimes, can become zeitgebers as well.

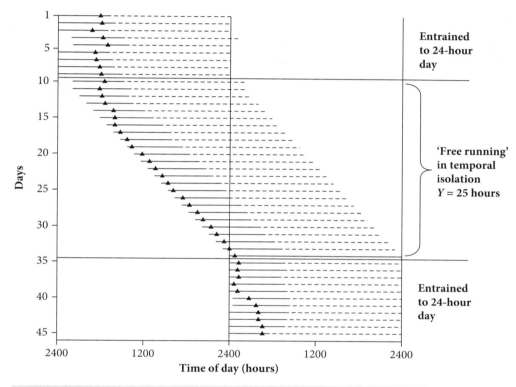

FIGURE 3.6 Free-running sleep–wake cycle of a volunteer in an isolation chamber with and without cues about the day–night cycle

To explore the effects of environmental cues on sleep patterns, volunteers were placed into an underground bunker set up in a cave where the temperature was constant and all external cues for day and night were removed. After a period of acclimation that included social interactions, meals at normal times and temporal cues (clocks, radio, TV), all cues that would indicate the time of day were removed, allowing observation of the 'free-running' cycle based on the participants' internal clock. In the absence of temporal (time-based) cues, participants went to bed and woke progressively later each day (about 10 o'clock the first 'night', about 11 o'clock the second 'night', about 12 o'clock the third 'night', and so on), eventually drifting from a 24-hour cycle into a 25 to 26-hour cycle. These cycles still depict circadian rhythms (as they still approximate one day). After a designated number of weeks, the temporal cues were reinstated, bringing their rhythms back into line with external cues.

Gamma-aminobutyric acid (GABA) has long been regarded as a sleep-promoting neurotransmitter, enabling the mind and body to relax, so we can fall asleep and sleep soundly throughout the night. The timing of GABA production and release are important for the modulation of circadian rhythms in the SCN. Many of the inhibitory pathways of melatonin synthesis and secretion use GABA as the neurotransmitter. Low GABA activity is linked to insomnia and disrupted sleep.

3.1.4 Sleep patterns across the life span

Figure 3.7 depicts the number of hours spent sleeping as a person ages and the percentage of total sleep time spent in REM sleep.

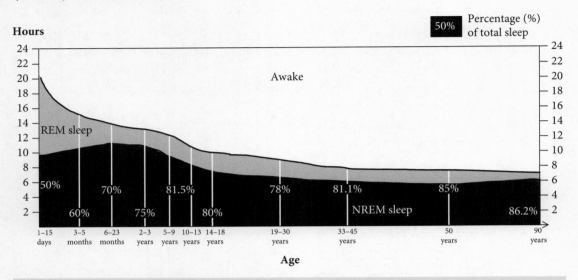

FIGURE 3.7 A graph of amount of sleep time a person has over the life span

As people age, they require less total sleep per 24-hour period. Newborns sleep almost all the time – up to 17–18 hours. Infants cycle through many sleep periods throughout the day. As they develop, they sleep longer at night and have fewer sleep periods during the day. Ten year olds usually sleep for 10–11 hours, adolescents should sleep for 9–10 hours, adults should sleep for 7–8 hours, and elderly people sleep for approximately 6 hours with little deep (NREM 3) sleep. Thus, there is a strong negative correlation between total hours slept per night and age.

The percentage of total sleep time spent in REM sleep also lessens as a person gets older. For example, a newborn spends approximately 50% of total sleep time in REM sleep compared with an adult, who spends approximately 20% of total sleep time in REM sleep. The fact that infants show a higher proportion of time in REM sleep is seen as supporting the theory that REM sleep aids in brain development and the formation of synapses.

3.2 Importance of sleep to mental wellbeing

3.2.1 Sleep deprivation

Sleep deprivation occurs through lack of sleep, which can lead to a variety of physiological and psychological effects, especially in relation to **affective (emotional) functioning**, **behavioural functioning** and **cognitive functioning**.

Symptoms of sleep deprivation

The typical effects of a short period of sleep deprivation include lethargy, irritability, headaches, loss of concentration, inattention, difficulty completing low-level boring tasks, inefficiency, confusion or misperception. When allowed to go to sleep, sleep-deprived individuals fall asleep more quickly and sleep longer than usual.

TABLE 3.6 Physiological and psychological effects of sleep deprivation

Physiological effects of sleep deprivation	Psychological effects of sleep deprivation
• Headaches • Dizziness • Drooping eyelids • Problems focusing the eyes • Lethargy, lack of energy and strength • Feelings of fatigue and sleepiness • Shaky hands (poor fine motor control) • Slower physical reflexes/reaction times for motor tasks • Slurred speech • Ataxia (physical strength and coordination deteriorate) • Impaired functioning of the immune system • Heightened sensitivity to pain/aches and pains in body • Autonomic levels (heart rate/respiration) drop/slow down • Drop in body temperature • Impaired production of certain hormones • Desynchronisation of biological rhythms • Microsleeps • REM rebound	• Memory impairment/lapses • Attentional difficulties, shorter attention span • Difficulty concentrating on simple or boring tasks (better with complex tasks) • Pronounced inefficiency • Difficulty processing information • Impaired creativity • Poor reasoning and decision-making / impaired judgement and risk-taking behaviour • Confusion or misperception • Cognitive disturbances (including disorientation, irrational/illogical thinking, possible paranoia and/or delusions) • Affective disturbance (including anxiety/hypertension) and increased emotionality (irritability and mood swings) • Apathy, lack of motivation *If prolonged:* • **Sleep disorders**, such as circadian rhythm disorders, narcolepsy and hypersomnia • Possibility of sleep deprivation psychosis (including hallucinations)

Performance of complex intellectual tasks is not significantly impaired by sleep deprivation. However, an individual's ability to follow simple routines is affected, as is their vigilance, attentiveness and mood. Low-level boring tasks are the most likely to be affected, as sleep deprivation may affect motivation rather than ability.

The symptoms of severe sleep deprivation are all temporary. Once allowed to catch up on sleep by getting a few hours extra over the days subsequent to their deprivation, individuals display no long-term harmful effects from such sleep loss.

REM rebound occurs after being selectively deprived of REM sleep or due to **partial sleep deprivation**. In the nights following REM sleep deprivation, people spend more time than usual in REM sleep, indicating a need to make up for lost REM sleep.

Total sleep deprivation is difficult to measure because after three or four sleep-free days, individuals involuntarily drift into **microsleeps**, which are short periods of drowsiness or sleep that intrude into the waking state. Microsleep episodes last from a few seconds to several minutes. While in a microsleep, a person fails to respond to outside information and, afterwards, is often not aware that a microsleep has occurred. Microsleeps increase with cumulative sleep debt. The more sleep deprived a person is, the greater the chance a microsleep episode will occur.

Effects of one night of full sleep deprivation compared to effects of alcohol

When controlling a motor vehicle, lack of sleep may be as dangerous if not more dangerous, than the effects of alcohol. The Centre for Sleep Research at the University of Adelaide published the findings of a study by Lamond and Dawson (1998) into performance impairment of drivers affected by fatigue. For years, the police have been warning the public that 'fatigue kills!' The study, which tested 60 volunteers over a year, has confirmed the police warning.

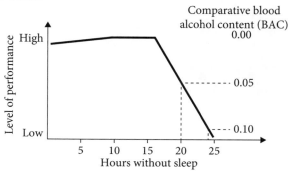

FIGURE 3.8 Effects of lack of sleep on performance compared to effects of blood alcohol content

The study reported that when participants were deprived of one night's sleep, they performed as badly in mental and physical tasks as when they were twice the legal blood alcohol limit for drivers in most Australian states (Figure 3.8). Volunteers who stayed awake for 20 hours produced performance levels equivalent to when they had a **blood alcohol concentration (BAC)** of 0.05%. In those who stayed awake for about 24 hours, the performance level was equivalent to when they had a BAC of 0.10% (twice the legal driving limit).

These results suggest that modest levels of sleep deprivation result in performances comparable to or worse than those seen at levels of alcohol intoxication considered to be unacceptable when driving, operating and/or working with dangerous equipment.

Participants tested when both drunk and sleep deprived are considered to be in an altered state of consciousness, where their **cognitive processes** are significantly impaired. Performances on speed and accuracy tasks while in an altered state of consciousness are generally poor, with decreased alertness, slower reaction times, poorer concentration, irrational/illogical thinking, difficulty making decisions, reduced ability to divide attention between competing stimuli, and so on.

A series of such studies spanning more than a decade at the Centre for Sleep Research have obtained consistent results, attesting to the reliability and validity of the findings, and providing additional support for the conclusions.

An important point is that fatigue is not just a lack of sleep or a consequence of long working hours. There are many factors that must be taken into account, such as the quality of sleep, which hours are worked in a 24-hour day and if sleep or work is in split shifts. There is a much greater chance of an accident occurring in the early hours of the morning due to the effects of sleep deprivation.

3.2.2 Circadian rhythm disorders

While the isolation studies (discussed earlier in section 3.1.3) demonstrated that our endogenous (internal) clock runs on a roughly 25-hour cycle, as diurnal animals most of our activity is during **daylight**, which means that the human body is constantly adjusting to Earth's 24-hour cycle as defined by the rising and setting of the Sun. However, gradual and sudden environmental changes can disrupt our biological rhythms, putting our internal body clock out of alignment with the Sun's cycle and causing changes to the sleep–wake cycle known as **circadian phase disorder**.

Exposure to light late in the night and early in the morning leads to phase advances that cause activity to start earlier the next day, whereas exposure to light in the evening and early night delays the circadian cycle, causing activity to start later the next day.

Delayed Sleep Phase Syndrome

Delayed Sleep Phase Syndrome (DSPS) involves an inability to reset the sleep–wake cycle in response to environmental time cues, resulting in a misalignment of the timing of sleep, peak period of alertness, core body temperature and hormonal and other daily rhythms relative to societal norms.

Because of the delayed release of sleep-inducing hormones (melatonin), a person's sleep is delayed by 2 or more hours beyond what is considered a typical sleep schedule. Those with DSPS are often described as 'night owls' who are often most alert, productive and creative late at night. If they go to bed at a 'normal' hour, they can lie awake in bed for a long time, unable to fall asleep until some hours after midnight. Unlike most other insomniacs, however, they fall asleep at about the same time every night, no matter what time they go to bed.

Duration and quality of sleep may be maintained if not disturbed, but the delayed onset of sleep causes difficulty waking up at the desired time in the morning; for example, to get up for school, work or social responsibilities. As a result, people with DSPS often suffer from lack of sleep, which can lead to daytime drowsiness or fatigue that can interfere with their ability to maintain relationships or perform at school or work.

The exact cause is unknown, but may be related to genetics, underlying physical conditions and a person's behaviour. This syndrome is especially common in adolescents and young adults, with a prevalence rate of 7–16%.

Many adolescents try to compensate by sleeping, on average, nearly 2 hours longer on weekends, and/or by taking long naps during the day. To some extent this may provide relief from daytime drowsiness, but irregular sleep schedules also pose problems by perpetuating the late sleep phase and disrupting the normal sleep cycle.

Advanced Sleep Phase Disorder

Advanced Sleep Phase Disorder (ASPD) is a syndrome in which the timing of sleep and the peak period of alertness occur several hours earlier relative to societal expectations regarding a normal circadian rhythm.

People with ASPD generally have difficulty staying awake unless they go to bed very early (usually between 6 p.m. and 9 p.m.), and often complain of sleep-maintenance insomnia because of their early morning awakening (between 2 a.m. and 5 a.m.) and inability to fall back asleep. Duration and quality of sleep are usually unaffected, but individuals with ASPD may have short sleep duration if they delay their bedtime.

While people with ASPD do not report the syndrome affecting their daytime work or school activities, being sleepy in the late afternoon or early evening can interfere with social responsibilities. The most common difficulty described by those with this condition is the effect on family and social relationships as evening activities are restricted by their need to sleep at a much earlier time than others.

This disorder affects about 1% of people in middle and older age and has the same prevalence among men and women. Researchers have also found that there may be an inherited genetic factor as 40–50% of people suffering from ASPD are related to someone who has the syndrome.

As they may be too drowsy in the late afternoon and evening, individuals with ASPD should be prudent about driving. They also do not cope well with evening or night shifts.

Shift work disorder

In industrialised societies, many people undertake **shift work**, which involves employment that takes place on a schedule outside the traditional 9 a.m. to 5 p.m. day. This can involve evening or night shifts, early morning shifts and rotating shifts.

Night-shift workers face the difficult task of having to adjust their endogenous biological clock to fit with an inverted night–day lifestyle. They often drive home in morning daylight, making it harder to reset their biological clocks, and can often fall back into a day–night schedule on their days off to spend daytime with their family, further disrupting their circadian adjustments.

Night workers who try to go to bed during the middle of the day get little (or poor-quality) sleep as human biological clocks promote sleepiness in the early morning hours. This, combined with subsequent fatigue due to chronic partial sleep deprivation, most probably accounts for the peak in job-performance errors, fatal traffic accidents, and engineering and industrial disasters between midnight and 6 a.m. On-the-job sleepiness is a major concern among night-time long-distance transport workers (including truck, bus and train drivers and airline crews) and medical staff (doctors and nurses).

Shift workers often work alternating rosters of day and night shifts, which makes it harder for them to realign their sleep–wake cycle to match their constantly changing work schedule.

One method of minimising the effects of shift work on sleep is to employ a rotating roster of shifts. A forward rotating schedule takes advantage of the body's free-running circadian rhythms. When work shifts change, it is easier to extend the 'waking day' than to compress it.

Bright light therapy

Light is the strongest entraining agent for synchronising circadian rhythms, and timed exposure to bright light is often used in the treatment of circadian phase disorders to influence the timing of the release of melatonin.

For entrainment purposes, bright room light over time may be sufficient; however, the use of a **bright light therapy** 'box' or lamp that emits very bright levels of light, much stronger than normal indoor ambient light, is often necessary to accomplish acute phase shifts. Special light visors and light glasses can also be worn.

Appropriately timed bright light exposure and avoidance of light at the wrong time of the day have been shown to be effective strategies to accelerate entrainment (or resetting) of circadian rhythms. The time of day when the light is used will depend on the disorder it is meant to correct.

Effective treatment of DSPS aims to realign circadian rhythms with the desired sleep and wake schedule, including adherence to good sleep hygiene, avoidance of bright light in the evening and increasing light exposure in the morning. Exposure to bright light in the evening may help in cases of ASPD.

For night-shift workers, circadian rhythms need to be delayed, so light therapy is applied in the early evening and night-time hours to attempt to offset the release of melatonin in order for the highest sleep propensity to occur during the day, rather than at night.

3.2.3 Improving sleep–wake patterns

Adaptation to zeitgebers

The alternating sleep–wake cycle is partly controlled by our internal 'body clock', which synchronises many of our physiological processes according to a daily rhythm. As discussed earlier, these circadian rhythms are affected by external cues (zeitgebers). Some of these are environmental, such as the cycles of light and dark, which affect the activity of the SCN, while others occur as a part of our daily routine. Getting good-quality sleep requires maintaining and adhering to this physiological clock, rather than engaging in behaviours that work against it.

While we may not have full control over our circadian rhythm, there are methods that we can apply to try to better entrain our 24-hour sleep–wake cycles to improve the quality of our sleep and enhance our mental **wellbeing**.

Sleep hygiene

Sleep hygiene is the term used to describe healthy habits, behaviours and environmental factors that can contribute to having a good night's sleep and waking up well-rested. Establishing sustainable and beneficial routines enables quality sleep to feel more automatic.

Factors that promote good sleep hygiene include:
- maintaining a stable daytime routine
- exposure to sunlight
- appropriately timed exercise
- good eating and drinking patterns
- avoiding daytime napping
- keeping a stable sleep schedule
- avoiding technology (and blue light) at night
- following a relaxing sleep ritual before going to bed

- making your bedroom a comfortable sleeping environment (cool, dark, quiet and free of disruptions)
- going to bed when you are sleepy
- only using your bed for sleep.

Daily routines

Your behaviours during the day can affect how well you sleep, as our regular activities form aspects of the zeitgebers that regulate our daily cycle. Even if you didn't sleep well and are tired, it is important that you try to keep a stable daytime routine as planned to avoid disrupting your normal circadian rhythm, which may cause further sleep problems.

Adhering to a consistent sleep schedule with regular sleep–wake times (even on weekends and holidays) will help preserve your circadian rhythm, making it easier for you to fall asleep and wake up every day. Varying your morning wake-up time or bedtime can hinder your body's ability to adjust to a stable daily rhythm and could lead to a circadian shift. To help maintain your sleep–wake cycle, establish a daily routine with set activities that happen during the day and another set of activities that happen at night, so that you find yourself getting sleepy at about the same time every night.

Where possible, avoid daytime naps, or if required, keep them short and take them early in the afternoon to make sure that you are tired at bedtime. Long naps late in the afternoon or evening can push back your bedtime schedule and affect the quality of your night's sleep.

Proper, regular exercise, even as little as 10 minutes a day, can improve sleep quality and duration, as well as your overall health. Timing is important, however, and where possible, moderate exercise should occur earlier in the day as the morning light helps us to wake up. If you do exercise in the evening, it should finish 3–4 hours before bedtime as your body needs time to wind down. Exercise should be avoided in the hours leading up to bedtime, as physical activity at that time increases your energy levels and body temperature, which may make it harder to fall asleep.

Nightly routine

In addition to reinforcing our circadian rhythm through the timing of routines during the day, our nightly habits also determine how easily we can fall asleep. Calculate a target bedtime that allows you to get the right amount of sleep before your fixed wake-up time and do your best to be ready for bed around that time each night.

Following the same pattern in the 30–60 minutes before going to bed establishes a set of cues your brain associates with winding down, calming down and getting ready for sleep. Begin this routine by doing a relaxing activity until you start to feel tired, then finish preparing to go to bed by putting on your pyjamas and brushing your teeth.

Go to bed only when you actually feel tired or sleepy, otherwise you will only reinforce bad habits such as lying awake in bed. If you haven't been able to get to sleep after about 20 minutes or more, get up. Not being able to fall asleep may cause you to become frustrated, which can keep you awake even longer. Do something calming to help you unwind, such as reading a book, until you feel sleepy, then return to bed and try again. Avoid doing anything that is too stimulating or interesting, as this will wake you up even more.

Your bedroom should be an environment where you are able to relax and doze off. Good sleep is more likely if your bedroom feels tranquil, restful and free of disruptions. The room should be quiet, cool and dark enough to prevent light from interrupting your sleep. Your mattress and pillow should be comfortable and provide you with the correct level of support.

Try not to use your bed for anything other than sleeping. If you treat your bed like a second lounge room for reading, working, talking to friends on the phone or watching TV, your mind will link your bedroom with activity and mental stimulation. It is best to avoid doing these things to strengthen your brain's association between your bed and sleep, thereby making it easier to fall asleep.

Light (including daylight and blue light)

People's circadian rhythms or cycles are influenced by the amount and intensity of light to which they are exposed. This effect occurs because light alters the amount of the hormone melatonin produced by the pineal gland.

Research suggests that spending 30–45 minutes each day outdoors in natural sunlight, especially early in the day, helps to boost wakefulness and entrain an appropriate circadian rhythm, which will then foster better-quality sleep at night.

As a part of our nightly preparation for going to bed, we should decrease our exposure to bright light in the evening and during the night by dimming indoor lighting for 1–2 hours before bedtime to help promote melatonin levels, thereby encouraging sleepiness.

The bedroom should also be kept as dark as possible to assist with the release of melatonin to encourage the onset of sleep.

We should also avoid bright artificial light from TVs, mobile phones and computer screens in the lead-up to bed. The **blue light** from technology acts like sunlight, lowering your melatonin levels and making it harder to fall asleep. Such devices also cause mental stimulation that keeps your brain alert and which is hard to shut off. Experts advise that we should limit the use of such devices at night, especially in the 30–60 minutes before going to bed, and definitely keep them out of the bedroom.

Temperature

Our body temperature follows a daily cycle, whereby it rises during waking hours and drops when we sleep.

Falling asleep is easiest if your core temperature is falling or low (evening and night), and most difficult if it is high or rising (morning and afternoon). Therefore, the ambient temperature of our sleep environment should be at a cool, but comfortable level, ideally 16–18°C, with a quilt that is not too heavy and hot or light and cold.

Having a hot bath 1–2 hours before bedtime can also help in this regard. It initially raises your body temperature, then helps you feel sleepy as your body temperature drops back down again.

Eating and drinking patterns

Studies have found that changing our **eating and drinking patterns** can affect our circadian rhythm. Altering the timing of food intake has the potential to cause misalignment of metabolic processes, making you feel alert or tired at different times than those you've become accustomed to. Therefore, in order to maintain an appropriate cycle, we should endeavour to preserve a regular meal schedule.

Research also suggests a healthy, balanced diet with reduced sugar and carbohydrates and increased fibre and protein may improve sleep quality. Other recommendations that may assist with getting to sleep include avoiding stimulants after 2 p.m. and not eating or drinking caffeine or alcohol for at least 2–3 hours before bedtime.

Eating late, especially if it is a big or spicy meal, can mean you're still digesting when it is time for bed, which would make it difficult to go to sleep. If, on the other hand, having an empty stomach at bedtime is distracting, a light snack may be in order. Several sources suggest drinking a warm glass of milk; this contains tryptophan, which acts as a natural sleep inducer.

Glossary

Advanced Sleep Phase Disorder (ASPD) A circadian rhythm sleep disorder characterised by a sleep pattern that is significantly earlier (by at least 2 hours) than a conventional or socially desirable sleep pattern, resulting in evening sleepiness and early-morning insomnia (inability to sleep). It may lead to impairments in social and/or occupational functioning, and is more common in elderly people

affective functioning A way of assessing levels of consciousness by seeing if an individual is aware of their emotional state and are showing a normal range of appropriate emotions

alpha waves The brain wave pattern seen in an *electroencephalogram (EEG)* that is typical of normal resting wakefulness, usually with eyes closed, or when a person is practising a meditative state of awareness, characterised by moderately high amplitude and moderate frequency

altered state of consciousness (ASC) A psychological state that is characteristically different from normal waking consciousness, including altered levels of self-awareness, perceptions, emotions, sense of reality, orientation in time or space, responsiveness to stimuli, and memorability; includes sleep, and may be drug-induced

amplitude In relation to brain waves, this is the height, or size, of the peaks and troughs indicating the level of voltage within the electrical impulses

awareness A subjective condition of being cognisant of something, from internal states or feelings to external, environmental stimuli

behavioural functioning A way of assessing levels of consciousness by observing an individual's actions and comparing these to established baselines

beta waves Brain waves characteristic of normal waking consciousness, with a low amplitude and high frequency

blood alcohol concentration (BAC) The amount of alcohol present in the bloodstream

blue light The bright artificial light emitted by technology such as TVs, mobile phones and computer screens

brain waves The electrical discharges of the living brain as recorded by an EEG. Brain waves are measured and described in terms of their amplitude (voltage) and frequency (cycles per second)

A+ DIGITAL FLASHCARDS Revise this topic's key terms and concepts by scanning the QR code or typing the URL into your browser.

https://get.ga/aplus-vce-psych-u34

bright light therapy A treatment for circadian rhythm phase disorders that exposes people to intense but safe amounts of artificial light for a specific and regular length of time to help synchronise their sleep–wake cycle with a normal external day–night cycle

circadian phase (or circadian rhythm sleep) disorders Any sleep disorder caused by a mismatch between a person's internal circadian rhythm and their actual or required sleep schedule, including jet lag, shift work sleep disorder, advanced sleep phase disorder and delayed sleep phase disorder. Referred to in the *Diagnostic and Statistical Manual of Mental Disorders* (DSM-5) as circadian rhythm sleep–wake disorder

circadian rhythm Any regular, automatic variation in physiological or behavioural activity that repeats at approximately 24-hour intervals, including the *sleep–wake cycle* and body temperature

cognitive functioning A way of assessing levels of consciousness by evaluating an individual's ability to process and/or recall information and comparing this to established baselines

cognitive functions Processes that involve mental activity in order to process information, such as thinking, reasoning, planning, learning, memory, perception, using language and so on

consciousness A person's (or animal's) internal and external awareness, including awareness of sensations, perceptions, emotions and thoughts

consciousness continuum A model that represents the different levels of consciousness, spanning from heightened awareness with extreme concentration, to our normal waking consciousness, through to lowered levels such as daydreaming and sleep, down to a complete lack of awareness

daylight Natural sunlight, which acts as the primary zeitgeber to boost wakefulness during the day and entrain an appropriate circadian rhythm

deep sleep A general term given to describe the lowest level of NREM sleep dominated by very slow, large-amplitude delta brain waves where the body is very relaxed and the person is unresponsive to any environmental stimuli and very difficult to wake up

Delayed Sleep Phase Syndrome (DSPS)
A *circadian rhythm sleep disorder* characterised by a sleep pattern that is significantly later (by at least 2 hours) than conventional sleep patterns, resulting in later sleep onset and wake times. It causes impaired alertness and performance during the day and is common in adolescence. It is also referred to as delayed sleep phase disorder (DSPD) or delayed sleep–wake phase sleep disorder

delta waves The lowest-frequency brain waves with high regular amplitude, characteristic of the deepest stages of non-REM sleep (stage 3)

eating and drinking patterns The routines in timing of food intake that act as zeitgebers for metabolic processes within our circadian cycle

electroencephalography (EEG) The process that detects, amplifies and records electrical activity of the brain via a series of electrodes placed at various points on the surface of the participant's scalp

electromyography (EMG) The process that detects, amplifies and records the electrical activity of muscles, generally via electrodes attached to the jaw or lower face, in order to measure muscle tension

electro-oculography (EOG) The process that measures eye movements or eye positions by detecting, amplifying and recording electrical activity in the muscles around the eyes via electrodes attached to the upper face, usually one above each eye and one just to the side of each eye, near the temple

entrain The process of determining or modifying the phase or period of circadian cycles

frequency In relation to brain waves, the number of brain waves per second

hypnagogic state A state when alpha waves begin to present on the EEG and a person is drifting from wakefulness to sleep

hypnic jerk A reflex muscle spasm throughout the body that occurs when a person is falling asleep (NREM stage 1)

hypnopompic state The final stage of the sleep cycle leading out of sleep as the individual wakes up

induced state of consciousness Different levels of awareness along the consciousness continuum that have been purposefully generated or brought about by other factors (e.g. alcohol, medication/drugs) or due to external influences (e.g. meditation, hypnosis)

K complex A short burst of high-amplitude brain waves, experienced in stage 2 NREM sleep

light The main environmental cue that influences the sleep–wake cycle, making humans awake and alert during the day and sleepy when it is dark

melatonin A hormone secreted by the pineal gland that causes drowsiness and helps to regulate the sleep–wake cycle

microsleep Episodes of sleep lasting only a few seconds that are not detected by the brain, posing danger in situations such as driving. Microsleeps are characterised by momentary lack of awareness, sudden waking due to the head falling forward or body jerks, and occur because of sleep deprivation

naturally occurring state of consciousness Different levels of awareness along the consciousness continuum that normally occur as part of the natural daily cycle (e.g. daydreaming, sleep and dreaming)

normal waking consciousness (NWC) The state of awareness we experience during wakefulness when we are aware of our surroundings and engage effectively in daily work, learning and social experiences. It is characterised by low-amplitude, high-frequency irregular activity in an electroencephalogram

NREM (non-REM) sleep The stages of sleep not associated with rapid eye movements. These stages contain little dreaming and people can still move while in NREM sleep

paradoxical sleep A term for REM sleep where the brain is showing a high level of activity while the body is inactive and extremely relaxed (to the point of paralysis)

partial sleep deprivation Getting some sleep in a 24-hour period but less than normally required for optimal daytime functioning

physiological responses Objective measures of bodily functions that can be used to assess differing states of consciousness

pineal gland An endocrine organ (gland) located deep within the forebrain that secretes melatonin, which regulates body rhythms and the sleep–wake cycle

REM (rapid eye movement) sleep The sleep stage that occurs between stages of non-REM sleep in which most dreaming occurs, typically accounting for between one-quarter to one-fifth of total sleep time. It is characterised by high-frequency, low-amplitude electroencephalogram readings that resemble normal waking consciousness, accompanied by paralysis of skeletal muscles. It is also called paradoxical sleep because of the similarity in brain wave patterns with wakefulness

REM rebound A natural compensatory process that occurs after being deprived of REM sleep or after periods of stress in which a person experiences increased frequency, depth and intensity of the REM stage of sleep

shift work Hours of paid employment that are outside the period of a normal working day and may follow a different pattern in consecutive weeks

sleep An altered state of consciousness governed by circadian rhythms during which awareness of ourselves and the environment is suspended. It is characterised by a series of typical changes in sleep electroencephalogram readings, muscle tension, eye movements and other physiological changes that accompany the different stages of sleep

sleep cycles The ultradian NREM/REM cycles of about 90 minutes that are evident while individuals progress through their night's sleep

sleep deprivation The condition of having had insufficient sleep to support optimal daytime functioning

sleep diary A daily log that can be used to record an individual's sleep–wake pattern with related information, usually over several weeks

sleep disorder A persistent disturbance of typical sleep patterns (including the amount, quality and timing of sleep) or the chronic occurrence of abnormal events or behaviour during sleep

sleep hygiene A set of behavioural treatment techniques to manage insomnia (inability to fall asleep) and improve sleep patterns, which may include: using the bed only for sleeping and sex; reduced daytime napping; decreasing caffeine and avoiding it after 4 p.m.; a regular bedtime routine; and keeping a sleep diary

sleep inertia The 'groggy', confused disorientation and poor memory of sleep events experienced by individuals woken during NREM stage 3

sleep spindles A type of brain activity characterised by a short burst of high-frequency brain waves, experienced during stage 2 NREM sleep

sleep–wake cycle The natural biological rhythm that produces the pattern of alternating sleep and wakefulness over a 24-hour period

state of consciousness An individual's level of mental awareness of sensations, perceptions, memories and feelings, which can range from being fully awake and focused through to unconsciousness

suprachiasmatic nucleus (SCN) A cluster of neurons in the hypothalamus located directly above the optic chiasm that regulates the body's circadian rhythms, particularly the sleep–wake cycle, using information about the intensity and duration of light received from the retina via the optic nerve

temperature The ambient levels of heat or cold that can affect an individual's ability to fall asleep

theta waves The irregular brain wave pattern with a frequency between alpha and delta waves and a mixture of higher and lower amplitude. These brain waves are characteristic of a person in light sleep (NREM stages 1 and 2)

ultradian rhythms Biological cycles that occur more than once a day (e.g. the NREM/REM cycles of about 90 minutes within the sleep period)

video monitoring In sleep research, the use of video cameras to record externally observable or audible physiological responses of the individual during the night

wellbeing Overarching term that encompasses an individual's physical, emotional and mental health

zeitgeber An external cue such as light, temperature, noise or food that influences the activation or timing of a biological rhythm, such as the circadian sleep–wake cycle; from the German meaning 'time giver'

Revision summary

Use the following summary of syllabus dot points and key knowledge within Unit 4 Area of Study 1 to ensure that you have reviewed the content thoroughly. Provide a brief definition or comment for each item to demonstrate your understanding or code them using the traffic light system: green (all good), amber (needs some review) or red (priority area to review). Alternatively, write a follow-up strategy.

How does sleep affect mental processes and behaviour?	
The demand for sleep	
• Definition of consciousness	
• The consciousness continuum	
• Normal waking consciousness	
• Altered states of consciousness	
• Naturally occurring states of consciousness	
• Purposely induced states of consciousness	
• Sleep	
• Patterns within the sleep cycle	
– REM (rapid eye movement) sleep	
– NREM (non-REM) sleep	
• Measurement of responses associated with sleep	
– Sleep diaries	

»	– Video monitoring	
	• Physiological responses	
	– EEG (electroencephalography)	
	– EMG (electromyography)	
	– EOG (electro-oculography)	
	• The regulation of sleep–wake patterns by internal biological mechanisms	
	• Circadian rhythm	
	• Ultradian rhythms	
	• Zeitgebers	
	• Suprachiasmatic nucleus (SCN)	
	• Melatonin	
	• Pattern of sleep across the life span	
	– Total amount of sleep required at different ages	
	– Changes in the proportion of REM and NREM across the life span	
		»

Importance of sleep to mental wellbeing	
• The impact of partial sleep deprivation on psychological functioning	
• Comparison of the affective and cognitive effects of one night of full sleep deprivation to blood alcohol concentration readings of 0.05 and 0.10	
• Circadian rhythm phase disorders	
– Delayed Sleep Phase Syndrome (DSPS)	
– Advanced Sleep Phase Disorder (ASPD)	
– Shift work disorder	
• Treatment of circadian rhythm phase disorders through bright light therapy	
• Improving sleep hygiene	
• Adaptation to zeitgebers to improve sleep–wake patterns and mental wellbeing	
– Light (including daylight and blue light)	
– Temperature	
– Eating and drinking patterns	

Exam practice

The demand for sleep

Answers start on page 241.

Multiple-choice questions

Question 1

The term 'consciousness' is best defined as

A a state focusing on our internal existence and activities.

B the awareness of external stimuli and of one's internal state.

C a condition distinguished by specific physiological processes.

D the ability of a person to respond to objects and events within their external environment.

Question 2

Consciousness can be described as

A the link between mind and body.

B a stable, consistent pattern of thoughts and feelings.

C an ever-changing stream of awareness.

D being difficult to study because its processes are largely unknown to the individual.

Question 3

While attempting the questions for this chapter, you would be in a state of normal waking consciousness. Identify which of the following characteristics would **not** be consistent with such a state.

A Attention is focused on external stimuli or on internal thoughts and feelings.

B Thinking is clear and structured.

C Continuity of one's memory over time may be disrupted.

D There is an awareness of time constraints.

Question 4

Rebekkah is working quietly in her study period. During this time, she reminisces about the party she went to on the weekend, thinks about how to balance her part-time job with her basketball roster and tries to plan what she is going to wear to her Year 12 formal. Rebekkah's thought processes are most indicative of which state of consciousness?

A State of drifting fantasies

B An altered state of consciousness

C A highly focused state of awareness

D A state of normal waking consciousness

Question 5 ©VCAA 2018 SA Q24

Archer went to a sleep laboratory after reporting a history of insomnia and difficulty forming and consolidating memories. The sleep scientist gave Archer a prescription drug to make him feel relaxed enough to sleep, altering his state of consciousness.

Archer's physiological responses were monitored in three separate areas, as shown in the image on the right.

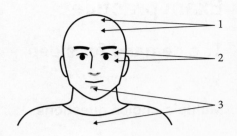

Which of the following identifies the equipment used to capture Archer's physiological responses at the points labelled 1–3 in the image?

	1	2	3
A	EMG	EEG	EOG
B	EOG	EMG	EEG
C	EEG	EMG	EOG
D	EEG	EOG	EMG

Question 6 ©VCAA 2017 SA Q32

Barry volunteered to stay overnight at a sleep laboratory so that his sleep patterns could be studied.

Which of the following identifies the qualitative and quantitative measures that could be used to indicate Barry's state of consciousness?

	Qualitative measures	Quantitative measures
A	electro-oculograph (EOG) electroencephalograph (EEG)	sleep diary video monitoring
B	electromyograph (EMG) video monitoring	electroencephalograph (EEG) electro-oculograph (EOG)
C	sleep diary video monitoring	electromyograph (EMG) electro-oculograph (EOG)
D	sleep diary electro-oculograph (EOG)	video monitoring electroencephalograph (EEG)

Question 7

Davin is experiencing difficulties getting enough good-quality sleep. He decides to consult a sleep clinic to try to find out what can be done.

As part of the process, Davin has to spend a night at the clinic, attached to several machines that monitor and record physiological changes throughout the night.

What data about Davin do these devices provide for the sleep clinicians?

	EEG	EMG	EOG
A	Brain wave patterns	Patterns of eye movement	Muscle tone and activity
B	Patterns of eye movement	Brain wave patterns	Muscle tone and activity
C	Muscle tone and activity	Brain wave patterns	Patterns of eye movement
D	Brain wave patterns	Muscle tone and activity	Patterns of eye movement

Question 8

The brain waves characteristic of an ordinary individual in an alert state of normal waking consciousness are

A alpha waves.
B beta waves.
C delta waves.
D theta waves.

Question 9

Recordings of brain waves in sleeping patients indicate that during the sleep cycle

A the brain changes in its pattern of activity.
B the brain shuts down, showing almost no activity at all.
C the brain experiences a general inhibition and reduction in its functioning.
D brain activity is no different from that recorded in subjects who are wide-awake.

Question 10

The electrical activity of our brain changes as we progress from being awake and alert through to a state of deep sleep. The sequence of the dominant brain waves within this progression is

A alpha, beta, gamma, delta.
B alpha, beta, delta, theta.
C alpha, beta, theta, delta.
D beta, alpha, theta, delta.

Question 11

A researcher in a sleep laboratory is most likely to infer that a healthy adult is in REM sleep if

A the electro-oculograph (EOG) shows high activity and the electromyograph (EMG) shows low activity.
B both the electroencephalograph (EEG) and the electro-oculograph (EOG) show low activity.
C the electroencephalograph (EEG) shows low activity and the electromyograph (EMG) shows high activity.
D both the electroencephalograph (EEG) and the electromyograph (EMG) show high activity.

Question 12 ©VCAA 2010 E1 SA Q37 (ADAPTED)

A typical night's sleep for an adult includes four to five sleep cycles.

Which of the following patterns best describes a typical sleep cycle from early in the night?

A Awake, NREM stage 3, NREM stage 2, NREM stage 1, REM, NREM stage 3, NREM stage 2, NREM stage 1
B Awake, NREM stage 1, NREM stage 2, NREM stage 3, REM, NREM stage 1, NREM stage 2, NREM stage 3
C Awake, REM, NREM stage 3, NREM stage 2, NREM stage 1, NREM stage 2, NREM stage 3, REM
D Awake, NREM stage 1, NREM stage 2, NREM stage 3, NREM stage 2, NREM stage 1, REM

Use the following information to answer Questions 13 and 14.

Eli is a healthy young adult who eats nutritious food and does not drink coffee after midday.

Question 13

Which of the following statements is most likely true of Eli's sleep pattern across a typical night?

A Periods of NREM stage 3 sleep occur more often as the night progresses.
B About 80% of the night is spent in REM sleep.
C Each cycle of sleep lasts for approximately 60 minutes.
D Periods of REM sleep become longer and occur more frequently as the night progresses.

Question 14

Within a normal night's sleep for Eli, NREM sleep

A becomes deeper within each successive cycle throughout the night.

B is spread evenly throughout the sleep cycle.

C comprises about 60% of time spent asleep.

D helps to restore bodily processes and repair tissue damage.

Question 15

Which of the following has **not** been suggested as a likely function of REM sleep?

A To enable stimulation needed for the brain to grow and develop

B To enable the restoration of chemicals needed to repair brain tissue

C To enable a long period for the nervous system to 'shut down' and rest

D To enable events that occurred during the day to be processed further to allow consolidation into long-term memory

Question 16

Explanations of the differences between REM and NREM have focused on the need for sleep in terms of

A bodily processes and tissue damage being repaired during the paralysis of REM sleep, while brain processes are restored during the slower brain activity in NREM sleep.

B children needing less sleep, especially REM, in comparison with adults because of their high energy levels.

C longer periods of sleep (especially NREM stage 3) occurring after heavy physical exercise.

D adults requiring longer periods of sleep in comparison with children in order to recover from balancing their work and social life.

Question 17

Circadian rhythms are biological cycles that

A are approximately 24 hours in duration.

B conform to a seasonal cycle.

C last for about 90 minutes.

D rotate on a monthly basis.

Question 18

In experiments conducted to find out whether people have an internal 'clock' that regulates their behaviour, human participants were deprived of time cues and placed into an underground bunker in a cave where there was no change in temperature or noise level.

Results of these studies found that the participants

A showed no alteration to the physiological processes associated with their normal circadian rhythms.

B slept significantly more than usual.

C adapted their physiological processes to a regular 25-hour cycle.

D slept significantly less than usual.

Question 19

The brain structure that responds directly to the amount and intensity of light and subsequently plays a role in regulating the sleep–wake cycle is the

A pineal gland.
B suprachiasmatic nucleus.
C hypothalamus.
D basal ganglia.

Question 20

The hormone melatonin, secreted by the pineal gland, seems to be a key factor in regulating

A consciousness.
B levels of arousal in response to a perceived threat.
C body temperature.
D the sleep–wake cycle.

Question 21

Which statement is accurate concerning REM and NREM sleep?

A An individual always cycles through the three levels of NREM before going into REM sleep.
B As we progress through the life span from birth to old age, the percentage of time spent in REM sleep decreases.
C The first half of the sleep cycle is dominated by the NREM stages, whereas the second half of the sleep cycle is overtaken by more frequent, progressively shorter periods of REM sleep.
D As their bodies do not require the restorative functions of the NREM cycle, children spend most of their sleeping time in REM.

Question 22

In terms of total amount of sleep and proportion of REM and NREM, which of the following sets of data **best** describes the changes in typical patterns of sleep across the life span?

	Age group	Number of hours slept	% of time spent in NREM sleep	% of time spent in REM sleep
A	Infancy (0–6 months)	10	80	20
	Adolescence (13–18 years)	6	70	30
	Old age (>85 years)	9	90	10
B	Infancy (0–6 months)	11	30	70
	Adolescence (13–18 years)	7	20	80
	Old age (>85 years)	5	14	86
C	Infancy (0–6 months)	20	50	50
	Adolescence (13–18 years)	6	60	40
	Old age (>85 years)	10	70	30
D	Infancy (0–6 months)	13	70	30
	Adolescence (13–18 years)	9	80	20
	Old age (>85 years)	6	85	15

Question 23 ©VCAA 2017 SAMPLE EXAM SA Q41

The following graphs show the typical sleep cycles for two distinct age groups.

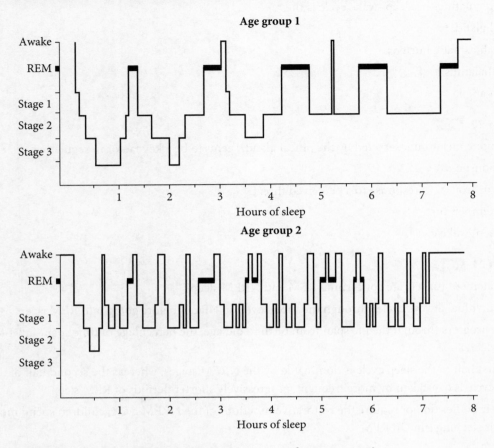

Whose typical sleep cycles are represented by Age group 1 and Age group 2?

	Age group 1	Age group 2
A	Infant	Adolescent
B	Infant	Elderly person
C	Adolescent	Elderly person
D	Elderly person	Infant

Short-answer questions

Question 24 (7 marks)

a Define 'consciousness'. — 2 marks

b Clarify what is meant by a 'state of consciousness'. — 2 marks

c Describe three factors that would indicate you are currently in a state of normal waking consciousness. — 3 marks

Question 25 (5 marks)

a Identify which state of consciousness people consider as providing them with a perception of reality. Explain the reasoning behind your answer. — 2 marks

b Outline how performance on cognitive tasks might indicate differing levels of consciousness. — 3 marks

Question 26 (8 marks)

a Describe three factors that would indicate an individual was experiencing an altered state of consciousness.   3 marks

b Clarify the difference between naturally occurring states of consciousness and those that have been induced.   2 marks

c Sleep is considered to be a naturally occurring altered state of consciousness. Identify three psychological characteristics that would distinguish sleep from normal waking consciousness.   3 marks

Question 27 (6 marks)

a Explain why researchers would prefer techniques that measure physiological responses to self-report methods in order to assess an individual's state of consciousness.   2 marks

b Describe two disadvantages of sleep diaries as a method for gathering data on sleep patterns.   2 marks

c Explain how video monitoring would be used in sleep research to indicate that the person is in REM sleep.   2 marks

Question 28 (4 marks)

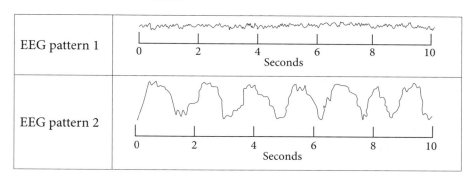

The two different patterns shown above are examples of electrical activity recorded from the brain during sleep.

a What is the term used to describe how 'big' the waves of electrical activity are?   1 mark

b What is the term used to describe how many waves occur every second?   1 mark

c Identify the stage of sleep indicated by pattern 1.   1 mark

d Identify the stage of sleep indicated by pattern 2.   1 mark

Question 29 (3 marks)

Define

a electroencephalograph (EEG).   1 mark

b electromyograph (EMG).   1 mark

c electro-oculograph (EOG).   1 mark

Question 30 (5 marks)

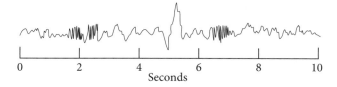

a On the basis of the brain wave pattern shown above, identify the stage of sleep that the person would be experiencing.   1 mark

b Name and describe the unique brain wave features that are characteristic of this stage of sleep.   4 marks

Question 31 (12 marks)

Using an electroencephalograph (EEG), an electromyograph (EMG) and an electro-oculargraph (EOG), researchers at a sleep laboratory record the physiological changes across the night's sleep for a group of healthy 30-year-old adults with no history of disease or injury to the brain.

For the two stages of sleep specified, complete the table below by describing the recordings obtained from each machine and clarifying what these recordings would indicate. 12 marks

Sleep stage	EEG recording	EMG recording	EOG recording
REM			
NREM stage 3			

Question 32 (7 marks)

a Describe the typical nightly sleep pattern for a young adult. 3 marks

b Outline some of the physiological attributes of REM sleep that distinguish it from the other sleep stages. 2 marks

c Summarise the relative proportion of time spent in NREM during a typical adult's nightly sleep cycle. 2 marks

Question 33 (4 marks) ©VCAA 2018 SB Q5 (ADAPTED)

a The figure below is a hypnogram representing the sleep cycle of a healthy adult.

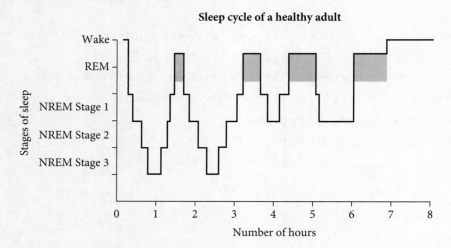

Sleep cycle of a healthy adult

Outline **two** differences between rapid eye movement (REM) sleep and non-rapid eye movement (NREM) sleep evident in the hypnogram above. 2 marks

b Compare how REM and NREM sleep would differ in a hypnogram of a healthy adolescent and a hypnogram of an elderly person. 2 marks

Question 34 (8 marks)

a Describe the changes that occur in sleep patterns across the life span from infancy through adulthood to old age, making reference to the average time spent sleeping and the amount of REM sleep at each age. 4 marks

b Based on studies across the life span, distinguish between the apparent function of REM sleep as opposed to that of NREM sleep. 4 marks

Question 35 (3 marks)

Describe the role that each of the following play in the regulation of sleep–wake patterns.

a Suprachiasmatic nucleus 1 mark

b Melatonin 1 mark

c GABA 1 mark

Importance of sleep to mental wellbeing

Answers start on page 247.

Multiple-choice questions

Question 1

Mild sleep deprivation is most likely to result in an increase in

A attention.

B irritability.

C concentration.

D energy levels.

Question 2

Studies have shown that participants with partial sleep deprivation will experience

A hallucinations and delusions.

B an enhancement in their memory for the affected period.

C a changed sleeping pattern involving a subsequent decrease in the amount of REM sleep.

D difficulty concentrating and performing simple tasks.

Question 3

Studies into the effects of one night of full sleep deprivation have found that

A after sleep deprivation, participants found it difficult to execute complex tasks but were able to do simpler tasks successfully.

B there was no effect on the participants' ability to perform coordination tasks.

C after sleep deprivation, participants experienced difficulty in performing simple tasks but were able to execute more complex tasks successfully.

D some of the participants experienced noticeable side effects for several weeks following the experiment.

Question 4 ©VCAA 2017 SA Q22

Frank had suffered one night of full sleep deprivation. Charlotte had consumed several alcoholic drinks and her blood alcohol concentration (BAC) was 0.10.

Compared to a legal BAC of 0.05, both Frank and Charlotte would be expected to show

A about the same cognitive impairment.

B better decision-making ability.

C no change to concentration.

D less exaggerated mood.

Use the following information to answer Questions 5–7.

Phoenix and her friends drove to a music festival in Byron Bay. They arrived the night before the festival and they all slept in Phoenix's small car for the night. Phoenix and her friends all experienced very disturbed sleep.

Question 5 ©VCAA 2019 SA Q29

What behavioural effect may Phoenix and her friends experience the next day due to being partially sleep deprived?

A An inability to sit still while listening to the music

B A lack of interest in making conversation with each other

C Being unable to remember the names of all the bands that they were listening to

D Feeling particularly hungry and wanting to visit a food truck for burgers and chips

Question 6 ©VCAA 2019 SA Q30

At one point, Phoenix was unable to remember the hair colour of the lead singer of her favourite band. Sleep deprivation is likely to contribute to her poor memory because

A sleep deprivation can result in poor cognitive functioning.

B affective functioning is compromised by sleep deprivation.

C music festivals have a compounding effect on sleep deprivation.

D the hallucinatory effects of sleep deprivation will cause memory problems.

Question 7 ©VCAA 2019 SA Q31

Phoenix and her friends stayed up watching bands all night. The next morning, Phoenix wanted to drive her car home, despite having not slept for the entire previous night or day. Her friends urged her not to drive because of the effect that sleep deprivation may have on her concentration.

The most accurate information to support the concern of Phoenix's friends is that a full night's sleep deprivation is equivalent to a blood alcohol concentration (BAC) of

A 0.10 and her eyelids might droop.

B 0.05 and she might have slower reaction times.

C 0.10 and she might not stay within her lane on the road.

D 0.05 and she might find it difficult to maintain the speed at which she is travelling.

Question 8

Our biological rhythms will **not** be disrupted by

A exposure to blue light from technology in the evening.

B shift work.

C working in an outdoor situation.

D staying up throughout the night to complete a work requirement.

Question 9

Delayed Sleep Phase Syndrome is typically caused by

A overproduction of the hormone melatonin.

B premature release of the hormone melatonin.

C too much exposure to sunlight during the day.

D environmental factors and delayed release of the hormone melatonin.

Question 10

People with Advanced Sleep Phase Disorder

A often complain of sleep-maintenance insomnia.

B experience a sudden, excessive and uncontrollable need to fall asleep at various times throughout the daytime.

C need to sleep for 12 hours or more per day.

D do not report the syndrome affecting their daytime work or evening social activities.

Question 11

In comparison to those who work during normal business hours, shift workers get

A more, but poorer-quality, sleep.

B less, but better-quality, sleep.

C more, and better-quality, sleep.

D less, and poorer-quality, sleep.

Question 12

One method of minimising the effects of shift work is to employ a rotating roster of shifts.

The biological rhythms of shift workers adapt more quickly when

A morning shifts are followed by afternoon shifts and then night shifts.

B shifts change constantly.

C night shifts are followed by afternoon shifts and then morning shifts.

D morning shifts are followed by night shifts and then afternoon shifts.

Question 13

Circadian rhythms have been shown to be influenced by

A internal mechanisms known as zeitgebers.

B our eating and drinking patterns.

C levels of the hormone adrenaline.

D levels of blue light from technology during the daytime.

Question 14

The amount and intensity of light to which an individual is exposed can affect the sleep–wake cycle by altering the levels of the

A hormone adrenaline produced by the adrenal gland.

B hormone dopamine produced by the substantia nigra.

C neurotransmitter noradrenaline released within the nervous system.

D hormone melatonin produced by the pineal gland.

Short-answer questions

Question 15 (5 marks)
a List three of the typical effects of a short period of sleep deprivation. — 3 marks
b Describe the symptoms of severe sleep deprivation. — 2 marks

Question 16 (5 marks)
a According to the research studies by Lamond and Dawson, how do the effects of one night of full sleep deprivation compare with those of blood alcohol concentrations? — 3 marks
b How can task performance be impaired by sleep deprivation? — 2 marks

Question 17 (6 marks)
a Explain 'circadian phase disorder'. — 1 mark
b Distinguish between Advanced Sleep Phase Disorder and Delayed Sleep Phase Syndrome. — 2 marks
c Describe the circadian phase difficulties experienced by night-shift workers. — 1 mark
d Explain how bright light therapy can be used in the treatment of circadian phase disorders. — 2 marks

Question 18 (5 marks) ©VCAA 2015 SB Q6a,c

Ernie started a new job working regular night shifts in a factory. In his new job he is required to operate a machine.

a When sleeping during the day, Ernie experienced an increase in the number of memorable and vivid dreams compared to when he was sleeping at night.

With reference to the characteristics and patterns of sleep, explain why Ernie experienced this increase in dreams. — 3 marks

b Ernie persevered with his new job for 6 months but found it difficult to cope with working night shifts for extended periods. He finally started looking for another job as he was worried that he might eventually become involved in an accident.

State **one** physiological effect of long-term sleep deprivation and identify **one** reason why it may increase the likelihood of Ernie injuring himself or someone else at work. — 2 marks

Question 19 (10 marks)
a Define 'zeitgeber'. — 2 marks
b Explain how each of the following can act as a zeitgeber and suggest how each needs to be considered to improve an individual's sleep–wake cycle.
  i Light — 2 marks
  ii Blue light — 2 marks
  iii Temperature — 2 marks
  iv Eating and drinking patterns — 2 marks

Chapter 4
Area of Study 2: What influences mental wellbeing?

Area of Study summary

This area of study applies the biopsychosocial model to acknowledge that mental health, like physical health, should be considered along a continuum that would change according to a variety of factors that could affect an individual's levels of functioning and their social and emotional wellbeing. The interaction of biological, psychological and social influences is demonstrated through an examination of the development and treatment of specific phobias, along with an exploration of protective factors and cultural determinants that promote and improve mental wellbeing and resilience.

Area of Study 2 Outcome 2

On completion of this outcome, you should be able to:

- discuss the concept of mental wellbeing
- apply a biopsychosocial approach to explain the development and management of specific phobia
- discuss protective factors that contribute to the maintenance of mental wellbeing.

The key science skills demonstrated in this outcome are:

- analyse and evaluate data and investigation methods
- construct evidence-based arguments and draw conclusions
- analyse, evaluate and communicate scientific ideas
- comply with safety and ethical guidelines.

Adapted from *VCE Psychology Study Design (2023–2027)* © copyright 2022, Victorian Curriculum and Assessment Authority

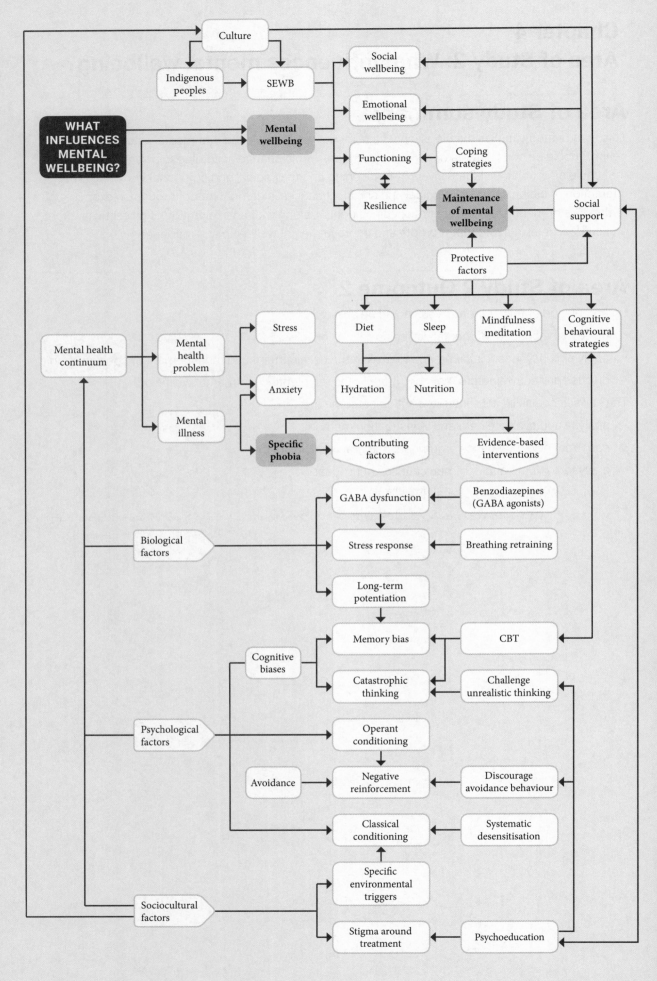

Summary of Units 1 & 2 prior knowledge

Concepts of normality

We all use and apply the concept 'normal'; however, its definition is often taken for granted. Generally, **normality** is considered to be what is acceptable or what can be expected to happen in most circumstances, whereas in a clinical context it is a condition that does not require treatment or assistance.

There is a fine and somewhat subjective line between normality and abnormality. Behaviours are defined as **abnormal** when they are statistically unusual, are not socially approved, cause distress to the person or interfere with an individual's ability to function. The definition of what constitutes 'abnormal' has varied across time and across cultures. Individuals also vary in what they regard as normal, abnormal or merely idiosyncratic behaviour. For example, a teacher who wears ties with cartoon figures on them may be seen as abnormal, as teachers usually wear plain ties. Or they may be viewed as merely expressing their individuality through their choice of apparel.

TABLE 4.1 Terms often used to describe 'normal' and 'abnormal' behaviour

Typical		Atypical
Behaviour that applies to, or is exhibited by, the majority of people and/or the majority of the time.	versus	Behaviour that applies to, or is exhibited by, the minority of people and/or is different from what is usually displayed.
Adaptive		**Maladaptive**
Behaviour that enables an individual to adjust to another type of behaviour or situation. Such behaviours are seen as a positive, constructive and productive response to a challenge.	versus	Behaviour that is often used to reduce one's anxiety, but the result is dysfunctional and non-productive. Such behaviours do not effectively address the actual problem and are often associated with various forms of mental illness.

TABLE 4.2 The main psychological approaches used to define normality

Model	Explanation of normality	Example
Biopsychosocial	A more eclectic, holistic approach incorporating psychological, biological and societal influences in order to define behaviour.	• Stress • Depression • Schizophrenia
Functional	Focuses on how effectively an individual is able to do what is expected in everyday life, taking into account the adaptiveness of the behaviour and how it affects the wellbeing of the individual or the social group.	• Addiction • Anxiety disorders • Stress • Mood disorders
Historical	As many behaviours evolve and change over time, the definition of normal behaviour involves an examination of the customs, habits and traditions within a particular society/culture to identify what is viewed as acceptable during a particular era.	• Smacking children was once permitted but is now discouraged. • Homosexuality was once classified as a mental disorder, but now it is not.
Medical	Describes normality in terms of physical health and its underlying physiological causes. Abnormal behaviour is seen as caused by an illness that usually has a physiological or biological cause. It can be diagnosed according to symptoms and signs of disorder, treated and, in most cases, cured by therapeutic methods within a hospital or clinical setting.	• Normal = healthy • Abnormal = sick/ill

TABLE 4.2 cont.

Model	Explanation of normality	Example
Situational	Normality is related to the factors to do with context (location or place) that are used to determine whether behaviour is typical or acceptable within that context.	• Wearing pyjamas is all right for going to bed, but not for attending school.
Sociocultural	Describes normality in terms of what a particular society or culture views as acceptable. Normality is based on influences of nationality, religion, ethnic group, peer group, family or other relevant groups. This approach identifies inappropriate behaviour so that social norms may prevail for the harmonious functioning within a particular society or culture.	• Wearing a kilt • Eating raw fish • Walking around naked
Statistical	Conclusions about normality are made on the basis of the analysis of numerical data and calculations as to the way that most people behave. Common measures are mean, median and mode, pertaining to the normal distribution curve.	• Height • Weight • IQ

Biopsychosocial approach to physical and mental health

Previously, the medical model of health and wellbeing considered mental health and physical health to be two separate concepts. The focus on the identification of objective symptoms within categorical systems based on the medical model reduces human beings to one-dimensional sources of data rather than encouraging practitioners to treat the whole person.

Critics of the medical model argue that psychological **disorders** may not reflect true psychopathology involving a deep internal problem, but instead are a poor adaptation to ordinary problems of living, often due to difficulty in the person's environment, in their current interpretation of events or in bad habits and poor social skills. It is important to observe patients over a period of time to understand their mental illness and its effects on their life.

The **biopsychosocial model** attempts to address this problem by providing a holistic approach that considers the individual as a unique being and their physical and mental health in terms of the dynamic interaction and integration of biological, psychological and social factors. This model suggests that these factors are all interlinked in promoting health and resilience or causing a vulnerability to disease. What affects the body will often affect the mind, and vice versa.

According to the biopsychosocial model, the state of being in good health is accompanied by good quality of life and strong relationships.

The biopsychosocial framework has contributed to application of a functional model of health and wellbeing. Diagnosis and treatment of illness focuses not only on the body but on the whole person in their social context, taking into account family and social support networks. The biopsychosocial clinician's task is to develop a broader understanding of disease processes by assessing the interrelationships of multiple systems and working with the patient to choose appropriate interventions, knowing that all systems will then be further affected.

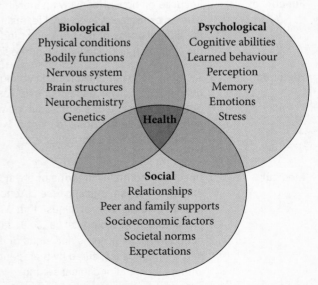

FIGURE 4.1 The biopsychosocial model of health

Factors that contribute to the development and progression of mental health disorders

A number of factors interact to contribute to the development and progression of mental health disorders.

The incidence and severity of a mental disorder often depends on the interaction of a number of key dynamics, beginning with underlying vulnerabilities, which may be brought out by life events and/or trauma, and prolonged by maladaptive behaviours and impaired reasoning. Protective factors (discussed later in section 4.3.1) comprise aspects that help to prevent or lessen the decline along the **mental health continuum** and may be strengthened and built on during treatment and recovery.

FIGURE 4.2 Factors that interact to contribute to the development and progression of mental health disorders

Each person may have certain attributes or be subject to influences that increase their susceptibility to potential behavioural health problems. **Predisposing risk factors** are those aspects within an individual's characteristics, many of them operating from their formative early years, that increase their vulnerability to other influences that act close to the time of the development and presentation of psychological problems. These factors play a role in the development of their personality and overall constitution.

Further to this, the individual may have characteristics, or be affected by circumstances, that make them vulnerable to the onset of potential behavioural health problems. **Precipitating risk factors** are the specific personal or situational dynamics that occur shortly before the development of a disorder. These appear to be the triggers for the occurrence or worsening of the psychological problems or behavioural response. Whether they produce a disorder at all, and what kind of disorder, depends partly on constitutional factors within the patient.

Perpetuating risk factors are longer term, ongoing factors that tend to maintain the psychological problem after it has been elicited, inhibiting recovery by prolonging the disorder and preventing its resolution. Sometimes a feature of a disorder makes it self-perpetuating (e.g. some ways of thinking commonly prolong anxiety disorders). Awareness of perpetuating factors is particularly important in planning treatment because they may be modifiable even when little can be done about predisposing and precipitating factors.

> **Note**
> An understanding of these factors will be necessary for certain concepts to do with phobias in section 4.2.

TABLE 4.3 Risk factors that could contribute to the development and progression of mental health disorders

	Biopsychosocial factors		
	Biological factors	**Psychological factors**	**Sociocultural factors**
Predisposing risk factors	• Genetic predisposition for vulnerability to specific disorders • Neurotransmitter dysfunction • Brain abnormality and/or dysfunction • Developmental disorders	• Personality traits (e.g. poor self-efficacy) • Sensitivity to stress and anxiety • Low self-esteem	• Parental modelling • Family history of mental health problems • Poverty and adversity
Precipitating risk factors	• Poor sleep • Substance use/misuse • Physical health problems • Head injury • Hormonal shifts	• Stress • Trauma	• Social stress • Loss of a significant relationship • Change in family and/or social dynamics • Loss of job/financial security
Perpetuating risk factors	• Long-term potentiation • Substance use/misuse to self-medicate	• Impaired reasoning and memory • Avoidance • Self-fulfilling prophecy to the 'sick' label	• The role of stigma as a barrier to accessing and receiving treatment • 'Sick' role (in response to attention and care received) • Family involvement and accommodation

Concept of cumulative risk

Risk factors do not *cause* illness. The presence of a risk factor does not necessarily mean that a person will have difficulties. However, risk factors tend to have a cumulative effect.

Cumulative risk refers to the compounded effects of a series or pattern of circumstances and events in an individual's life. It involves exposure to numerous biological, psychological and/or social risk factors that may be historical or ongoing, with the strong possibility of these factors being multiple, interrelated and coexisting over critical developmental periods. The continuous impact on the individual can be profound and exponential, covering multiple dimensions of their life.

Research examining cumulative risk has consistently found that the accumulation and interaction of risk factors exacerbates their effects, are predictive of poorer outcomes for the individual and significantly increase the likelihood of mental health problems. As a person encounters more risk factors, their susceptibility for developing a mental health disorder will incrementally intensify. What often differentiates high-risk individuals from lower-risk ones is the presence of multiple factors and/or adversities in their life histories.

Role of mental health workers

Because of the breadth and severity of various mental conditions, the question of what classifies as a mental disorder can be difficult to answer. To be diagnosed with a mental illness, a qualified mental health professional, such as a psychiatrist, psychologist or social worker, must evaluate the individual on the basis of what symptoms they have, how long the symptoms have persisted and how their lives are being affected. Most people affected by mental illness will be referred to a psychiatrist for further assessment and/or treatment.

A **psychiatrist** is a medical doctor who has received additional training in the study and treatment of mental illness and emotional disorders. A psychiatrist evaluates a person's mental health along with their physical condition and can prescribe medications.

Clinical psychologists specialise in working with people in the prevention, **diagnosis** or treatment of behavioural or mental health problems. While they are trained in the use of various psychological tests that can support a diagnosis, they are not medical doctors and cannot prescribe medications as a part of treatment.

Psychiatrists in particular are interested in descriptive **psychopathology**, aiming to describe the *symptoms* and *syndromes* of mental illness. This is both for the diagnosis of individual patients (to see whether the patient's experience fits any pre-existing classification) and for the creation of diagnostic systems (such as the *Diagnostic and Statistical Manual of Mental Disorders – DSM-5*) that define exactly which signs and symptoms should make up a diagnosis, and how experiences and behaviours should be grouped in particular diagnoses (such as clinical depression or schizophrenia).

Based on the *medical model*, classification systems allow clinicians and researchers to:
- standardise the description and interpretation of mental disorders
- provide vocabulary and a clinical shorthand to facilitate communication between professionals
- predict the **prognosis** (future course)
- consider appropriate treatment.

Classification systems also encourage research into the *aetiology* (cause or origin) of mental disorders and can serve as an education tool for teaching psychopathology.

Social stigma as a barrier to accessing treatment

While diagnostic labels may facilitate mental health professionals' communications and research, they can also create preconceptions that bias our perceptions of people's past and present behaviour. This was highlighted by Rosenhan's 1973 experiment in 12 mental hospitals. He sent in eight pseudo-patients who, upon admission, stated that they heard voices. After admission, they stopped saying they heard voices. Rosenhan found that the professionals within the mental hospitals didn't detect the fake patients and, in some cases, diagnosed them as having schizophrenia. Interestingly, some genuine patients suspected the pseudo-patients of being fake.

Labelling of disorders according to a medical model may have a detrimental effect on an individual's self-esteem and could become self-fulfilling prophecies, whereby the patient takes on the sick role and behaves in accordance with the perceived symptoms implied by the label.

These dynamics often contribute to the maintenance of unfair stereotypes attached to mental conditions. The resulting **social stigma** involves negative attitudes and beliefs held in the wider community and society in general that cause people to fear, reject, avoid and discriminate against those with a mental disorder, just for having an illness.

Because of the stigma associated with mental illness, many people do not seek out help, despite the availability of very well-researched and successful treatments. The perception that they are a weak person makes the individual feel worse about themself and prevents them from seeking help to avoid potential ridicule, humiliation and shame. For many, this dynamic can be a perpetuating factor that leads to a maladaptive cycle where the mental illness can continue and even strengthen.

People with a mental illness need the same understanding and support that is given to people with a physical illness. A deeper understanding of the nature and causes of such disorders can lead to removal of the stigma associated with mental illness, reduce generally negative depictions of people suffering from **mental disorders** and enable people to recognise symptoms and seek professional help.

4.1 Defining mental wellbeing

4.1.1 Mental wellbeing

Mental health is:
> a state of wellbeing in which the individual realises his or her own abilities, can cope with the normal stresses of life, can work productively and fruitfully, and is able to make a contribution to his or her community.
>
> World Health Organization (2001) *The World Health Report 2001: Mental Health*. New Understanding, New Hope.

The concept of resilience provides a framework for understanding the varied ways in which some individuals do well in the face of adversity rather than letting difficulties or failure overcome them. **Resilience** is the capacity to cope and deal constructively with change or challenges and bounce back to maintain or re-establish social and emotional wellbeing in the face of difficult events. Resilience enables people to swing back up the mental health continuum towards good **mental wellbeing**.

Resilience is not only dependent on the characteristics of the individual, such as a positive attitude and an ability to regulate emotions, but is greatly influenced by processes and interactions arising from family, friends and the wider environment, collectively discussed later in the chapter in relation to protective factors.

Depending on how risk and protective factors interact with each other, individuals may be resilient to some kinds of environmental stressors or outcomes but not others. Resilience can also change over time because of subsequent experiences and influences. Resilience provides an opportunity to grow, enhance our performance and develop better ways of dealing with life's demands. Therefore, resilience should not be viewed as a static, enduring ability, but rather as positive adaptation over time.

Recent discussions regarding resilience have shifted from an emphasis on factors or variables to focus on processes and mechanisms that individuals used to deal with the challenges they face, including ways of thinking and **coping skills** that enhance their ability to change and adapt.

4.1.2 Social and emotional wellbeing

Social and emotional wellbeing (SEWB) is a term preferred by many Aboriginal and Torres Strait Islander peoples to 'mental wellbeing' because it provides a holistic understanding of life and health that is consistent with Aboriginal and Torres Strait Islander peoples' holistic world views. SEWB is a holistic, multidimensional concept of health that includes mental health as one of seven domains of connection. The idea is that the wellbeing of an individual is influenced by the strength of their connections to the seven domains. Figure 4.3 shows the seven domains of connection as: 1. body and behaviours; 2. mind and emotions (i.e. mental health); 3. family and kinship; 4. community; 5. culture; 6. Country and land; and 7. spirituality and Ancestors. The figure also illustrates the important role of historical, social, cultural and political contexts as factors that influence SEWB. These broader contexts are referred to as 'determinants' of SEWB because they strongly influence the contexts into which Aboriginal and Torres Strait Islander children are born. Although the seven domains are named and described separately within the model, the dotted lines between them indicates their interconnectedness.

By connecting the individual to the seven domains and to the wider contextual determinants of SEWB, the physical and mental wellbeing of individuals influences and is influenced by the wellbeing of families and communities. Furthermore, the physical and mental wellbeing of individuals, families and communities influences, and is influenced by, their connections to culture, Country and spiritual Ancestral knowledge. It is the interconnections between the domains and their interconnections with the broader determinants that makes SEWB a holistic concept. Positive experiences and expressions of SEWB are associated with strong connections to the seven domains and strong interconnections between domains. Positive SEWB supports resilience against the broader social, cultural, historical and political contexts that can challenge the SEWB of Aboriginal and Torres Strait Islander people.

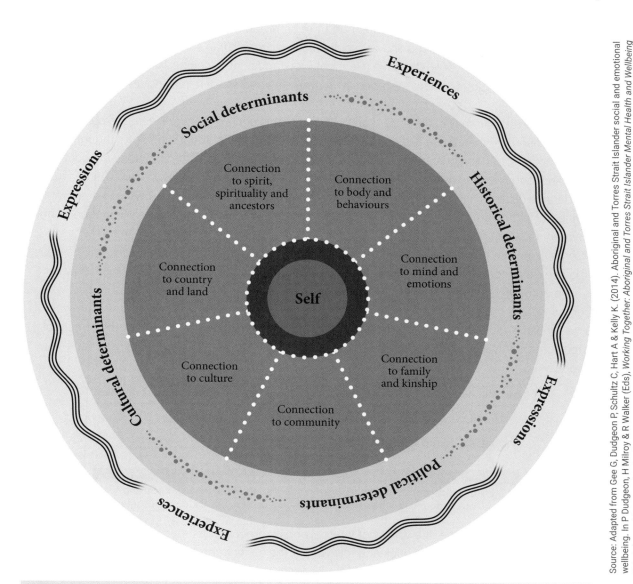

FIGURE 4.3 The model of Aboriginal and Torres Strait Islander peoples' social and emotional wellbeing

4.1.3 Mental health continuum

As with earlier discussions regarding consciousness, a person's health, both physical and mental, is not static, but will change over time along a *continuum* between two extremes, ranging from optimal wellness through to serious illness. The criteria used when deciding where an individual is placed along the continuum would most likely focus on the degree of severity of symptoms or level of impairment of **functioning**.

Healthy	Well	Ill	Sick	Disorder/dysfunction		(Death)
Healthy	Sniffles	Cough	Cold	Flu	Pneumonia	Pleurisy	(Death)

FIGURE 4.4 The health continuum

Mental health is a state of emotional and **social wellbeing** in which a person can form positive relationships with others, display high levels of functioning and think clearly and logically to fulfil their abilities, display resilience to cope with the normal stresses of life, work productively and be able to contribute to the community. With sound mental health, individuals feel good about themselves and are able to get on with their life smoothly.

Mental health problems, on the other hand, are common mental health complaints that affect our feelings, thoughts and actions and cause distress, and which can interfere with our performance and enjoyment in a range of life areas (e.g. school, work and relationships). They are usually understandable reactions to personal and social problems and are generally not too severe or long-lasting.

Although everyone may experience instances where this occurs, it is considered to be a **mental illness** or disorder when these feelings become so disturbing as to affect a person's thoughts, feelings, actions, perceptions and mental functioning (e.g. memory). Such conditions are usually more severe and longer lasting than mental health problems and become so overwhelming as to cause considerable distress and disruption to the person's capacity to function and cope with day-to-day activities, affecting their relationships with others, their ability to work and their participation in, and enjoyment of, life in general.

Some problems may be relatively common and temporary difficulties, such as anxiety or stress, or may be more complex and chronic, such as schizophrenia or bipolar disorder. Individuals who have a mental illness don't necessarily look like they are sick, especially if their illness is mild, whereas other individuals may show more explicit symptoms such as confusion, agitation or withdrawal. Mental illness affects about one in five people and almost everybody will encounter the effects of such disorders through a family member, close friend or workmate, or may even experience them personally.

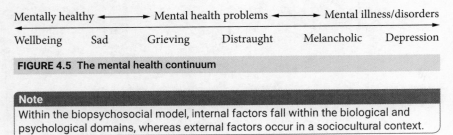

FIGURE 4.5 The mental health continuum

> **Note**
> Within the biopsychosocial model, internal factors fall within the biological and psychological domains, whereas external factors occur in a sociocultural context.

Because of the influence of a variety of internal and external factors that can fluctuate over time, an individual may progress back and forth along the continuum. Given the right treatment and support, a large number of conditions (both physical and mental) can be reversed back into health. While some disorders may be permanent, such as amputation, diabetes or schizophrenia, appropriate intervention can assist the individual to successfully manage their condition to regain an appropriate level of functioning and live a full and productive life.

TABLE 4.4 Mental health versus mental illness

Mental health	Mental illness
A state of emotional and social wellbeing	A state of emotional and social upheaval and/or disruption
A person can form positive relationships with others	The condition affects the person's ability to maintain effective relationships with others
A person can display high levels of functioning	The condition becomes so overwhelming as to cause disruption to the person's capacity to function
A person can think clearly and logically to fulfil their abilities	Feelings become so disturbing as to affect a person's thoughts, feelings, actions, perceptions and mental functioning (e.g. memory)
A person can display resilience to cope with the normal stresses of life	The condition causes considerable distress and inability to cope with day-to-day activities
A person can work productively and be able to contribute to the community	The condition affects their ability to work effectively
A person feels good about themself and is able to get on with their life smoothly	The person has difficulty participating in and deriving enjoyment from life in general

Stress

Stress involves prolonged levels of physiological *arousal* in response to threatening or challenging events.

The internal component of stress involves a set of neurological and physiological responses when confronted with a stressor. When the *fight–flight–freeze* response is triggered, the *sympathetic nervous system* is activated to prepare or mobilise the body to confront a situation or 'fight' it, or to flee from a situation (hence the term 'flight', such as when you are heading into an exam or when a dog suddenly barks at you). If the stressor is seen as overwhelming, the individual may experience the 'freeze' component, where they will stay still and may appear to have 'shut down'. This aspect would be associated with your mind going blank when asked a question in class.

The external component of the biopsychosocial model of stress involves a wide variety of psychosocial and environmental stimuli, *stressors*, which are perceived as physiologically or emotionally threatening and disrupt the body's *homeostasis (state of physiological balance)*.

> **Note**
> The concept of stress and its **contributing factors** were discussed in detail earlier in Chapter 1 (section 1.2).

Anxiety

Anxiety is a state of emotional arousal associated with feelings of worry or uneasiness that something is wrong or something bad is about to occur. While most people sometimes feel anxious, for some it is a source of extreme distress.

Anxiety disorders are diagnosed when the level of anxiety is out of proportion to the situation. Individuals with this disorder experience intense physiological arousal and tension, shaking, rapid breathing, increased heart rate, dizziness, feelings of losing control and general apprehension without an obvious reason or provocation. Such individuals often display maladaptive behaviours intended to reduce or avoid anxiety-producing stimuli.

Specific phobia

Fear is a natural, instinctive reaction to threatening situations. But people with *phobias* experience an intense, overwhelming irrational fear when exposed to a particular object or situation that ordinarily would not provoke such a reaction. A phobic response is so extreme that it causes significant distress and can interfere with everyday functioning.

> **Note**
> Phobias will be explored in detail in section 4.2.

Healthy	Mild effects		Moderate effects		Disorder/dysfunction
Eustress	Arousal	Tension	'Stress'	Moderate distress	Extreme distress
No anxiety	Nervousness	Agitation	Worry	**Anxiety**	Anxiety disorder	Panic
No fear	Fright (reaction)	Apprehension	Fear	Avoidance	**Phobia**	Terror

FIGURE 4.6 Applying the mental health continuum to stress, anxiety and phobia

TABLE 4.5 A comparison of stress, anxiety and phobia

Stress, anxiety and phobia are all:
• influenced by a range of biological, psychological and sociocultural factors • characterised by physiological changes, including the fight–flight–freeze response.

Stress	Anxiety	Phobia
The individual generally acknowledges the source as a specific stressor.	The individual is not always cognisant of the reason behind their anxiety.	The specific trigger for a phobic response is known by the individual.

Stress	Anxiety and phobia
This may be eustress or distress.	The response is always considered as causing distress.

Stress and anxiety	Phobia
In certain circumstances, this response would be thought of as a 'normal' reaction, as everybody has felt that way at some point.	Judged as being 'abnormal' in all circumstances.
Moderate levels can be adaptive and useful to motivate individuals towards a course of action.	Considered to be a maladaptive, hindering response.
These can be triggered by a wide range of stimuli.	By definition, a phobic response is triggered by a specific stimulus.
Avoidance of certain objects or situations could occur, but this is not necessarily the case.	Avoidance of the phobic stimulus is a central behavioural symptom.
If it is not handled effectively, it could have a detrimental effect on a person's functioning.	Drastically affects a person's functioning.
These can be risk factors that could lead to the development of a mental health disorder, especially if ignored and not dealt with.	A clinically recognised mental disorder.

4.2 Application of a biopsychosocial approach to explain specific phobia

4.2.1 Specific phobia

While it is natural to be afraid of certain things or situations, people with phobias experience intense, overwhelming fear that is outside of the norm when exposed to a specific stimulus. People with **specific phobia** are fine when the feared stimulus is not present. However, when faced with the object of their fear, they can become highly anxious and experience a panic attack. They are aware that their reactions are irrational and illogical, but the reactions are so powerful and unpredictable that they interfere with normal functioning. People affected by phobias can go to great lengths, drastically changing their lives to avoid situations that would force them to confront the thing that they fear. For example, an agoraphobe (who fears open spaces) may not leave their house, or someone with mysophobia (fear of germs) may not use a public toilet even if in desperate need.

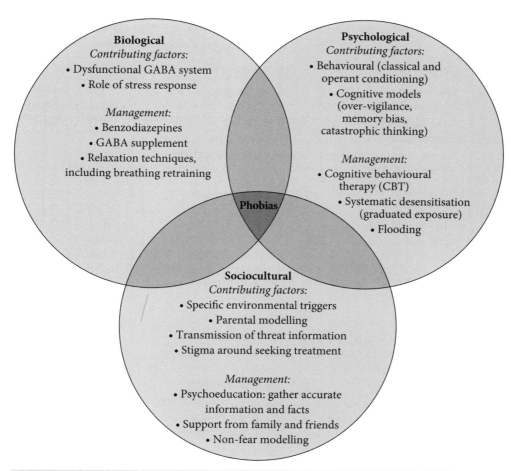

FIGURE 4.7 The biopsychosocial model of health applied to phobias

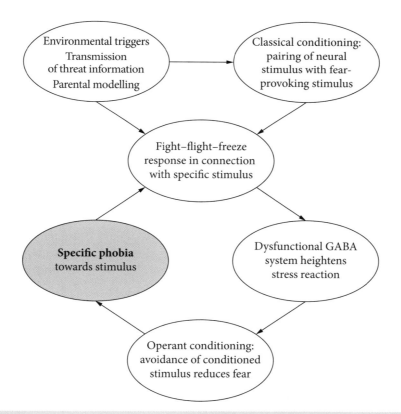

FIGURE 4.8 The interplay between the various factors within the biopsychosocial model that contribute to the development and progression of specific phobias

TABLE 4.6 Factors that could contribute to the development and progression of specific phobia

	Biopsychosocial factors		
	Biological factors	**Psychological factors**	**Sociocultural factors**
Predisposing risk factors	• Neurotransmitter **(gamma-aminobutyric acid [GABA]) dysfunction*** • Family history of mental health problems • Gender (more common in females)	• Developmental stage (specific phobias usually develop in childhood) • Personality traits, such as negativity and high inhibition	• Parental modelling • Family history of mental health problems
Precipitating risk factors	• The role of the stress response	• Behavioural models – classical conditioning whereby a traumatic event results in the association of fear with the phobic stimulus*	• **Specific environmental triggers*** • Transmission of threat information
Perpetuating risk factors	• Long-term potentiation* (constant pairing of fear with the object of that fear)	• Behavioural models*: operant conditioning – negative reinforcement through avoidance • Cognitive models*: cognitive bias, including memory bias and catastrophic thinking	• Stigma around seeking treatment* • Family involvement and accommodation towards the phobic behaviour
Protective factors	• Relaxation techniques including breathing retraining*	• Low sensitivity to stress • Cognitive behavioural strategies*	• Support from family, friends and community to challenge unrealistic or anxious thoughts and not encourage **avoidance behaviours***

*Listed in VCAA Psychology Study Design, so more likely to be in examination questions

> **Note**
> Factors in Table 4.6 and subsequent tables denoted with an asterisk (*) are listed in the VCAA Psychology Study Design, and are therefore more likely to feature in examination questions.

TABLE 4.7 Evidence-based interventions used in the treatment of specific phobias

Biological interventions	Psychological interventions	Social interventions

• The use of short-acting anti-anxiety benzodiazepine agents (GABA agonists) in the management of phobic anxiety*
 • Relaxation techniques*
 • Breathing retraining*
	• Psychotherapeutic treatments of phobia*
	 • The use of cognitive behavioural therapy (CBT)*
	 • Exposure therapies:*
	   – systematic desensitisation (graduated exposure)
	   – flooding
		• Psychoeducation for families/supporters to reduce maintenance of phobic behaviour*
		 • Challenging unrealistic or anxious thoughts*
		 • Not encouraging avoidance behaviours*

*Listed in VCAA Psychology Study Design, so more likely to be in examination questions

Biological contributing factors and their treatment

Long-term potentiation

Long-term potentiation is the neural basis of memory involving the long-lasting strengthening of synaptic connections of neurons, resulting in the enhanced functioning of the neurons whenever they are activated.

In the case of phobias, the repeated pairing of the fear response with the object of that fear causes stronger connections within the neural pathways associated with this behaviour, making it more likely that the individual will react in a fearful way to the specific stimulus.

Role of stress response

An aspect of natural selection suggests that evolution has made us biologically predisposed to acquire fearful responses through experience. As such, fear can serve an adaptive function by triggering the fight–flight–freeze response to make us avoid certain objects or situations that may threaten our survival. But because we are complex, thinking creatures living in cultural environments that present a variety of real and imagined threats, our fear can be triggered in circumstances in which it is not adaptive.

Further to this, the level of anxiety experienced by someone with a specific phobia tends to be excessive because their perception of threat is unreasonable and out of proportion to what it should be. Consequently, the physiological **stress response** they experience is often very severe and can persist at this high level for at least as long as the exposure or anticipated exposure to the phobic stimulus.

Relaxation techniques, including breathing retraining

Individuals can try to counter the effects of the stress response through the use of various *relaxation techniques*. These strategies employ any method, process, procedure or activity that helps a person to relax and attain a state of increased physiological and/or psychological calmness by relieving muscle tension and lowering autonomic levels, such as blood pressure and breathing rates, to reduce stress and anxiety. They may include deep breathing exercises, imagery, meditation and other techniques.

People with anxiety are thought to have abnormal breathing patterns. They may breathe faster and deeper than necessary (hyperventilation), resulting in an imbalance in the levels of oxygen and carbon dioxide in the blood, which may further increase the feelings of anxiety. **Breathing retraining** is a technique that enables those with anxiety to have more conscious control over their condition. It teaches correct breathing habits in order to slow down breathing by taking slow, deep inhalations and exhalations through the nose to prevent hyperventilation and restore the balance of oxygen and carbon dioxide. As well as helping to reduce physiological arousal associated with anxiety, breathing training may also help people feel as if they have more control of their anxiety. Breathing training can be used by itself or in combination with other treatments.

Dysfunction of neurotransmitter GABA

Because too much excitation can lead to irritability, restlessness, insomnia and seizures, it must be balanced with inhibition. Gamma-aminobutyric acid (GABA), the most abundant inhibitory neurotransmitter in the brain, reduces the activity of the central nervous system (CNS), acting like a 'brake' to the excitatory neurotransmitters (norepinephrine, adrenaline, dopamine and serotonin) during times of runaway stress and normally inhibiting the elevated physical responses to fear/anxiety.

Studies have shown a connection between decreased levels of GABA in the brain and anxiety symptoms, leading researchers to hypothesise that some people develop anxiety because they have a dysfunctional GABA system. Low levels of GABA in the brain of such individuals would be due to their failure to produce, release or receive sufficient quantities of GABA required to regulate neural activity in the brain. Lower levels of GABA mean that levels of arousal initiated by the excitatory neurotransmitters go unchecked, thereby leading to overstimulation and feelings of increased anxiety.

Use of short-acting anti-anxiety benzodiazepine agents (GABA agonists) in management of phobic anxiety

Anti-anxiety drugs, such as **benzodiazepine** agents, are often used in combination with psychotherapy and are commonly prescribed in the short term to calm the body by reducing physiological arousal and tension associated with the fight–flight–freeze response in order to help people cope with anxiety and panic attacks. They act as a **GABA agonist**, by imitating GABA's inhibitory effects on postsynaptic neurons throughout the brain to lower CNS activity and induce relaxation to calm the body and reduce arousal. The continued use of sedatives or mild tranquilisers can cause problems of dependence/addiction.

Inefficient processing of *serotonin* may also contribute to anxiety. *Selective serotonin reuptake inhibitor (SSRI)* drugs are also used commonly to treat anxiety disorders and depression. They raise the level of serotonin in the brain by preventing it from being reabsorbed back into the cells that released it. A recent study using functional brain imaging techniques suggests that the effects of SSRIs in alleviating anxiety may result from a direct action on GABA neurons.

Medication will not cure anxiety disorders, but can keep symptoms under control while a person receives psychological treatment.

Psychological contributing factors

Behavioural models

Behavioural models view anxiety disorders as a product of fear conditioning, stimulus generalisation, reinforcement of fearful behaviours and observational learning of other people's fear.

Precipitative factors lead to the development of phobias through **classical conditioning**, either directly or vicariously via observational learning. A stimulus that normally does not elicit fear suddenly takes on this role after being associated with a fearful episode that an individual has experienced personally, or second-hand through seeing or hearing about another's terrifying experience. The stimulus thereby becomes a conditioned stimulus (CS), leading to a conditioned response (CR) of fear.

Once acquired, phobias are maintained and perpetuated through **operant conditioning**. When a person begins to avoid the conditioned anxiety-producing stimulus, he or she experiences a decrease in anxiety, which **negatively reinforces** the behaviour and prevents extinction of the fear.

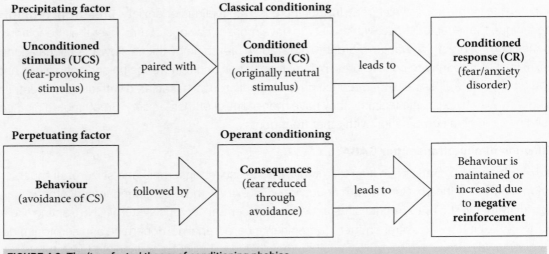

FIGURE 4.9 The 'two-factor' theory of conditioning phobias

Cognitive models

The cognitive models focus on distortions of thinking that cause people who suffer from phobias to show a **cognitive bias** through a pattern of inaccurate thinking, whereby they create their own subjective reality resulting in an illogical interpretation of and an inaccurate judgement about other people and situations. For someone with a phobia, the attentional problem of over-vigilance means that their perception of the phobic stimulus will always be distorted, chronically overestimating the severity of even minor danger cues to see threats in harmless or neutral situations and perceive the object as more threatening than it really is.

Their interpretation will be supported by **memory bias**, in that they will only remember negative or threatening experiences with the stimulus that will most likely be blown out of all proportion to what actually happened. **Catastrophic thinking** would then occur as they would only think about irrational worst-case outcomes, overestimating the potential dangers or negative implications of their experience with the phobic stimulus.

Those with phobias also tend to focus their attention internally, becoming more aware of bodily sensations, and are more likely to misinterpret those sensations as ominous. This thinking can lead to a vicious circle as; for example, an increased heart rate is misinterpreted as a sign of impending harm, leading to anxiety, which in turn leads to increased heart rate.

Psychotherapeutic treatment

Cognitive behavioural therapy (CBT)

Cognitive behavioural therapy (CBT) is a form of therapy that combines cognitive and behavioural therapies to redirect thoughts, feelings and behaviour, based on the idea that emotional or behavioural problems result from unrealistic or irrational thinking.

CBT helps people with phobias face their fears by teaching them new skills in order to react differently to the situations that trigger their phobia. Patients also learn to understand how their thinking patterns contribute to the symptoms, how to change their beliefs to reduce or stop these symptoms and, in time, how to accept whatever was causing their extreme anxiety.

By identifying irrational negative thinking along with the possible helpful interpretations of a situation, the cognitive component of CBT teaches sufferers how to change their thoughts (e.g. thinking rationally about whether there are any risks).

The behavioural component of CBT aims to alleviate symptoms by teaching sufferers how to change their reactions to anxiety-provoking situations (e.g. to engage in deep breathing).

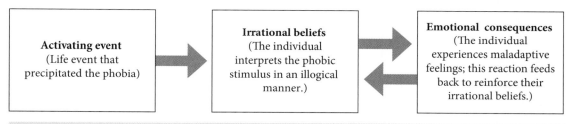

FIGURE 4.10 Factors influencing irrational beliefs

Behavioural therapies

From the behavioural perspective, fears are considered to be learned responses. A stimulus that can elicit an unconditioned fear response may be paired with another stimulus, which thereby becomes a conditioned stimulus capable of eliciting a conditioned fear response. According to the behaviourist, what can be learned can also be unlearned. If the conditioned stimulus is repeatedly presented without the unconditioned stimulus, extinction will occur.

Behavioural therapy may involve gradually exposing people to situations that trigger their anxiety. **Systematic desensitisation**, also known as **graduated exposure therapy**, includes a process by which individuals extinguish the association between the phobic stimulus and anxiety. First, the patients are taught relaxation techniques that they can use at each stage of the process. They are then told to visualise the feared conditioned stimulus until they are able

Situation	Fear rating
Holding a live beetle in your hand	100
Holding a dead beetle in your hand	85
Watching a beetle crawling across the table	80
Holding the jar with the live beetle	70
Looking at a live beetle in a jar	50
Holding a jar with a dead beetle	45
Holding a plastic toy beetle	40
Looking at a picture of beetles	25
Thinking about beetles	10

FIGURE 4.11 A fear hierarchy for a phobia related to beetles

to do so in a relaxed manner. The patient establishes a fear hierarchy which defines a series of tasks, graded in order of difficulty from 0 to 100 (e.g. imagine feared stimulus – picture – touch picture – actual stimulus – touch actual stimulus). To help clarify this for the patient, the hierarchy may be visually represented as a mountain or a ladder to show their progress. Then, step by step, incremental exposure allows the patient to progressively face the phobic stimulus, practising the relaxation technique in order to learn to cope with and overcome the fear associated with each step of the hierarchy until the phobia is eliminated. Higher-level exposures are not attempted until the patient's fear subsides for the lower-level exposure. The process aims to gradually extinguish the conditioned fear response and replace it with the specific relaxed response. Several studies indicate that desensitisation is the most effective and long-lasting treatment for a broad range of specific phobia.

Sociocultural contributing factors

Specific environmental triggers

An *environmental trigger* is a specific object, event and/or experience that can act as a stimulus for the onset of a particular condition.

Traumatic events can trigger anxiety disorders in some individuals. For example, individuals who have been bitten by a dog may become conditioned to fear dogs and develop a specific phobia. Phobias with a traumatic origin may develop acutely, having a more sudden onset than other phobias that develop gradually over time.

Parental modelling and transmission of threat information

Several studies show a strong correlation between parents' fears and those of their children. Although genetic factors play a role in some cases, some researchers believe that many children can learn fears and phobias through vicarious classical conditioning just by observing their parent's or other family member's fearful reaction to an event.

Transmission of threat information refers to the delivery of information from parents, other family members, peers, teachers, the media and other secondary sources about the potential threat or actual danger of a particular object or situation.

Stigma around seeking treatment

As phobias are based on anxieties and fears that are quite personal, other people can find it difficult to empathise. This might result in stigma and discrimination, which can make it much harder for people to speak openly about what they're going through and seek help, despite there being very well researched and successful treatments available. Further to this, the individual could feel ashamed about seeing a psychologist because people may see this as a sign of weakness, which could prevent them from seeking treatment.

The stigma attached to psychological disorders can lead the individual to apply defence mechanisms such as rationalisation, avoidance or denial. This can lead to a maladaptive cycle where the condition will continue and even strengthen, making it difficult for the individual to recognise their behaviour and seek help on their own. It often takes family members, friends or co-workers to persuade a person with a psychological problem to seek treatment. When family relationships are supportive and honest, this will often help to resolve issues and improve the ability of family members to cope with their problem.

Psychoeducation

Psychoeducation may be used to provide people, either individually or in a group, with information about their condition to assist in its management by increasing their knowledge and understanding of symptoms, recovery patterns and treatment options and services available.

In relation to phobias, psychoeducation would provide the patient, their families and supporters with information about the typical reactions and dynamics associated with the disorder. The aim is to help the patient to manage their condition by challenging unrealistic or anxious thoughts and focusing on adaptive behavioural change rather than encouraging and reinforcing avoidance behaviours.

4.3 Maintenance of mental wellbeing

4.3.1 Protective factors that maintain mental wellbeing

Individuals may have qualities or be exposed to influences that make them resilient in the face of potential behavioural health problems. **Protective factors** are those aspects, such as an individual's strengths, social supports and positive patterns of behaviour, that prevent or reduce the likelihood or severity of conditions due to the various risk factors present.

An awareness of protective factors is particularly important for a therapist in planning treatment because they may be modifiable, even when little can be done about predisposing and precipitating factors.

TABLE 4.8 Protective factors that maintain mental wellbeing

	Biopsychosocial factors		
	Biological factors	**Psychological factors**	**Sociocultural factors**
Protective factors	• Adequate diet* • Adequate sleep* • Exercise • Relaxation techniques • Medication (if necessary) • Reduced substance use	• Mindfulness meditation* • Cognitive behavioural strategies • Psychotherapeutic treatments • High resilience • Effective coping skills • High self-esteem • High self-efficacy • Personality traits: engaged, insightful	• Positive relationships*: support from caring family, friends and community • Stable family and social dynamics • Social competence • Cultural background • Psychoeducation

*Listed in VCAA Psychology Study Design

Impact of adequate diet, sleep and exercise on physical and mental health

Substantial research has demonstrated that physical health and mental health are critically linked. There appears to be a strong positive correlation between the two, with good physical health supporting and maintaining good mental health. If an individual feels well physically, they usually feel better mentally and will experience a better quality of life.

The reverse is also true. Poor physical health can lead to an increased risk of developing mental health problems. Similarly, poor mental health can negatively impact on physical health, leading to an increased risk of some conditions. It is therefore important that people with a mental illness or disorder receive good-quality physical as well as mental healthcare.

A poor **diet**, inadequate **sleep** and inactivity can be both causes and consequences of a variety of physical ailments, and consequently represent precipitating and/or perpetuating risk factors for a number of mental disorders.

A healthy lifestyle, on the other hand, is a protective factor that can make a big difference to your mental as well as physical health.

A nutritious diet is a crucial factor in influencing the way we feel, and is critical for healthy brain functioning. Eating healthy foods that contain lots of vitamins, minerals and antioxidants nourishes the brain and protects it from 'free radicals' produced when the body uses oxygen, which can damage cells. Good intestinal health also promotes the production of about 95% of our serotonin, the neurotransmitter and neuromodulator that mediates our moods and helps regulate sleep and appetite. Good sleep patterns improve concentration and energy levels and assist in faster recovery from illness.

As about 75% of brain tissue is water, adequate hydration is necessary for it to function properly and maintain our mental health. A steady supply of water keeps our blood flowing smoothly. Dehydration slows down our circulation, lowering blood flow, which means less oxygen, glucose and nutrients are supplied to the brain, altering how we think and feel. Even mild dehydration can quickly produce a headache, reduce our ability to concentrate and affect mood, increasing anxiety and the risk of depression. Higher levels of dehydration further impair cognitive function and can result in delirium. Severe dehydration can cause unconsciousness and even coma, finally leading to death.

As well as increasing physical fitness to effectively cope and deal with the physiological effects of stress, exercise can boost mood, concentration and alertness and may also improve other psychological outcomes, such as social competence, self-esteem, confidence and optimism. In addition to research that has shown regular exercisers to have better mental health and **emotional wellbeing** and lower rates of mental illness, longitudinal studies have demonstrated that physical exercise is a protective factor that seems to reduce the risk of developing mental disorders. Exercise programs are now an integral part of intervention strategies for a number of psychological disorders to aid recovery and prevent relapse.

Cognitive behavioural strategies

Cognitive approaches to wellbeing assume that our thoughts mediate between a stimulus and our emotional reactions to it. Problem behaviours and emotions therefore result from faulty beliefs and expectations about situations, rather than the situations themselves. Strategies aim to change maladaptive thought patterns by training people to restructure their thinking to view themselves in new, more positive ways.

As an extension of this approach, *cognitive behavioural strategies* attempt not only to teach people to think in more adaptive ways, but also to practise their new ways of thinking in everyday life.

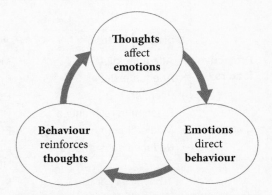

FIGURE 4.12 The cycle of thoughts, emotions and behaviour

These strategies aim to help the individual:

- recognise the connections between their thoughts, feelings and behaviour
- review how they managed certain life events
- consider all possible interpretations of situations
- monitor their negative thoughts
- challenge their negative thoughts with evidence
- substitute a more constructive, realistic explanatory style for any irrational interpretations
- learn a more positive approach to change their behaviour
- focus on practising behaviours that are constructive and productive.

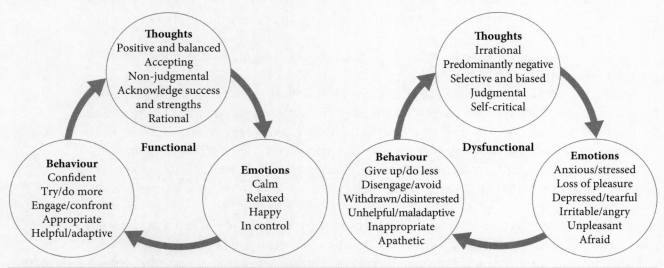

FIGURE 4.13 Comparison of functional and dysfunctional cycles of thoughts, emotions and behaviour

Mindfulness meditation

Meditation is a process employing mental exercises to achieve a highly focused state of consciousness. Many stress counsellors recommend meditation to promote relaxation and help to clear one's head to enable a clearer reappraisal of the situation.

In its simplest form, concentrative meditation involves assuming a relaxed sitting or lying position and breathing deeply, slowly and rhythmically. Attention is directed only at the breathing movements of the diaphragm, and all other thoughts are blocked from consciousness. This can generate a sense of relaxation and an inner, private reality as one proceeds into a naturally induced altered state of consciousness with increasing alpha and theta brain waves and decreasing heart rate, respiration rate, oxygen consumption and carbon dioxide expiration.

Meditation need not be thought of as formal mantra chanting, but simply an activity that provides a respite from upsetting thoughts.

Mindfulness meditation involves bringing awareness to the here and now with the goal of being fully present in the current moment and what you are currently experiencing, in order to centre your thoughts and avoid distractions. By focusing on the present, you are not stressing about what has happened in the past or what is to come in the future.

Mindfulness meditation can be as simple as taking a break and being silent, even for a few minutes. Breaks are essential to disconnect from the constant flow of stimulation in order to allow the mind and body to rest.

A common, evidence-based technique to deal with this is mindful breathing. In a similar way to the process of concentrative meditation, your breath functions as an anchor. By taking deep breaths and focusing on what is happening, feeling your chest expand and contract and listening to the sound of your breath, you can block out everything around you and bring your focus back to the present. This simple breathing technique not only clears your mind but can put you in a state of relaxation.

Applying mindfulness techniques can help maintain your mental and physical wellbeing by reducing stress, encouraging creativity by improving focus and concentration, and increasing resilience.

Social supports

A key component in maintaining our mental wellbeing involves our relationships with our family, friends and the community, not only to satisfy our need for contact with others, but to enable access to **social support** networks and resources.

Many studies show that having good social support correlates with better physical and mental health. Researchers believe that supportive social networks buffer the effects of stressful circumstances by providing a person with care and comfort, access to helpful resources and advice about how to evaluate and manage problems.

By providing emotional support during times of stress, these social networks can demonstrate concern and nurture the individual to let them know that they are not alone, and that help is available when it is required.

Social interaction can remove the person from the stress-producing situation, redirecting their focus away from the stressor and their negative emotional state, and put them into a setting where they can gain potential social support from others. Talking things through with someone else can often help us to reappraise the situation and provide alternative strategies to make things less stressful.

Cultural determinants

Aboriginal and Torres Strait Islander peoples' understandings of social and emotional wellbeing (SEWB) includes considering the influence of broader social, political, historical and cultural contexts as determinants of SEWB. These broader contextual factors present ongoing challenges to achieving positive SEWB for many Aboriginal and Torres Strait Islander people. For example, historical and ongoing government policies that affect the lives of many Aboriginal and Torres Strait Islander people have been directly linked to the over-representation of Aboriginal and Torres Strait Islanders in statistics measuring mental health issues, mortality rates, alcohol and substance abuse, high unemployment, poverty and contact with the justice system. The experiences of the Stolen Generations and the ongoing intergenerational impacts of the policies of forced removal of children and cultural assimilation clearly demonstrate how disrupting connections to family, communities, culture, Country and spiritual and ancestral knowledges negatively impacts SEWB.

We focus here on the influence of **cultural determinants** as powerful determinants of SEWB. The term cultural determinants refers to the extent to which broader cultural contexts enable Aboriginal and Torres Strait Islander people to learn and express their own unique cultures. Two of the strongest cultural determinants of SEWB are maintaining a sense of **cultural continuity** and the right to **self-determination**.

Cultural continuity can be defined as the process that enables cultural knowledge, values and practices to be transmitted over generations, and that integrates and connects individuals and communities with the past and future of their culture. Developing and maintaining a sense of cultural continuity is protective against the effects of racism. It engenders a sense of pride in being connected to the oldest living cultures on earth and gives a sense of hope for the future. Cultural continuity is developed and strengthened through the revitalisation of traditional languages and cultural practices, such as traditional methods of medicine and healing, and through being able to maintain connections to Country.

Self-determination is the fundamental right of Aboriginal and Torres Strait Islander peoples as the First Peoples of Australia to shape their own lives, so that they determine what it means to live well according to their own values and beliefs. Cultural continuity and self-determination are connected because cultural continuity depends on people being able to exercise their right to self-determination. Self-determination is demonstrated in the many community-led organisations that have been developed by and for Aboriginal and Torres Strait Islander communities (e.g. Aboriginal Community Controlled Health Organisations), in the development of approaches like the SEWB model that express Aboriginal and Torres Strait Islander world views, and in the ongoing activism of Aboriginal and Torres Strait Islander peoples to have a say in the policies that affect them. Self-determination has its most powerful current expression in seeking an Aboriginal and Torres Strait Islander voice to Parliament, articulated in the Uluru Statement from the Heart.

Being able to express cultural continuity and self-determination is protective against risk factors and increases the resilience and overall SEWB of Aboriginal and Torres Strait Islander people and communities.

Glossary

abnormal Behaviours that are statistically unusual, not socially approved, cause distress to the person or interfere with their ability to function

adaptive Positive, constructive and productive behaviours that enable an individual to adjust to a situation or respond to a challenge

anxiety An emotional state characterised by the anticipation of danger, dread or uneasiness as a response to an unclear or ambiguous threat

anxiety disorder A class of mental health disorders that feature feelings of fear, panic and the anticipation of danger, preventing normal functioning; accompanied by physical symptoms associated with threat, such as increased heart rate, muscle tension, sweating and rapid breathing

atypical Behaviour that is different to what is usually displayed by an individual and/or is displayed by the minority of people

avoidance (behaviours) Behaviours that attempt to prevent exposure to a fear-provoking object, activity or situation

behavioural models In relation to therapy, an approach to understanding and treating or managing a mental disorder that emphasises the role of learning and experience. The therapy focuses on extinguishing undesirable behaviour patterns through the application of classical and/or operant conditioning techniques

benzodiazepines A group of medications used in the short-term treatment of anxiety (also anxiolytics); as GABA agonists, they enhance the GABA-induced inhibition of overexcited neurotransmitters, calming nervous activity

biological factors Within the biopsychosocial framework, physiological influences that can affect an individual's wellbeing (e.g. genetics, brain function, general physical health)

biopsychosocial model An approach that proposes that health and illness outcomes are determined by the interaction and contribution of biological, psychological and social factors

breathing retraining The process of identifying incorrect breathing habits and replacing them with correct ones

catastrophic thinking When a person repeatedly overestimates the potential dangers and assumes the worst of an object or event

classical conditioning A form of associative learning in which a stimulus that naturally evokes a specific response is paired over a series of trials with a neutral stimulus that does not usually produce this response. An association occurs between this neutral stimulus and the unconditioned stimulus during this acquisition stage; after conditioning, this stimulus alone will elicit the response that it formerly did not produce

clinical psychologist Psychologists who specialise in working with people in the prevention, diagnosis or treatment of behavioural problems or mental health problems

cognitive behavioural therapy (CBT) An evidence-based psychological treatment approach that teaches clients to apply cognitive behavioural strategies to recognise and change negative and unproductive patterns of thinking and behaving. It is intended as a brief treatment program that can be effective within weeks to treat conditions such as anxiety, depression and other mental health problems

cognitive bias An automatic tendency or preference for processing or interpreting information in a particular way, producing systematic errors in thinking when making judgements or decisions

contributing factors Aspects of an individual's makeup or experience that are seen as having a causal connection in some way to their current condition

coping skills An individual's ability to employ their competencies and resources to manage their response to a stressful or unpleasant situation

cultural continuity The process that enables cultural knowledge, values and practices to be transmitted over generations, and that integrates and connects individuals and communities with the past and future of their culture

cultural determinants The extent to which broader cultural contexts enable Aboriginal and Torres Strait Islander peoples to learn and express their own unique cultures

cumulative risk The compounded effects of exposure to numerous risk factors within a series or pattern of circumstances and events in an individual's life

diagnosis The process of identifying and classifying an illness, disorder or abnormality on the basis of the presence of particular symptoms or results of tests and examinations

A+ DIGITAL FLASHCARDS Revise this topic's key terms and concepts by scanning the QR code or typing the URL into your browser.

https://get.ga/aplus-vce-psych-u34

diet An individual's pattern of food intake. Good diets provide sufficient nutrients, vitamins, minerals and antioxidants to nourish the body and brain to maintain overall health and wellbeing

disorder Any physical and/or psychological abnormality

emotional wellbeing An individual's ability to understand their emotions and manage them constructively to meet the demands of everyday life without experiencing too much distress

evidence-based interventions Different forms of therapy or treatment that have been shown to work in other cases as demonstrated by appropriate empirical research studies

functioning How effectively an individual is able to do what is expected in everyday life, taking into account the adaptiveness of the behaviour and how it affects the wellbeing of the individual or the social group

GABA agonist A chemical substance that binds to a *neuroreceptor* to produce a similar sedative effect to that of GABA to inhibit a *postsynaptic neuron* (e.g. *benzodiazepine* drugs)

GABA dysfunction It is hypothesised that some people develop anxiety because they fail to produce, release or receive a sufficient amount of GABA to regulate neuronal transmission in the brain, leading to overstimulation, and thus heightened anxiety

graduated exposure therapy A process used in the treatment of disorders involving fear and anxiety whereby an individual is gradually subjected, step by step, to increasingly similar stimuli to the conditioned stimulus itself through a series of tasks, to eventually extinguish the conditioned response

long-term potentiation A form of neural plasticity that results in a long-lasting strengthening of neural connections that form the basis of memory

maintenance of mental wellbeing Aspects including an individual's strengths, resilience, social supports and positive patterns of behaviour that enable them to continue in a state of mental health

maladaptive (behaviour) Non-productive and dysfunctional behaviour that is potentially harmful and prevents a person from meeting and adapting to the demands of everyday living

memory bias A tendency to remember information of one kind at the expense of another kind, including the bias towards remembering negative and threat-related experiences associated with specific phobia

mental disorder Psychological dysfunction that usually involves significant impairment of a person's thoughts, emotions or behaviours that causes distress to themselves or other people, and also affects their day-to-day functioning

mental health A state of emotional and social wellbeing in which a person can form positive relationships with others, display high levels of functioning and think clearly and logically to fulfil their abilities, display resilience to cope with the normal stresses of life, work productively and be able to contribute to the community

mental health continuum A model that acknowledges a person's mental wellbeing will change over time, ranging from optimal wellness through to mental health problems, to serious mental illness or disorder

mental health problem A disruption to how a person thinks, feels and behaves, but to a lesser extent than a mental health disorder

mental illness A health condition with symptoms that severely affect the way a person thinks, feels and acts, and which may cause them or other people distress, as well as causing difficulty in functioning or coping with everyday life; often involves behaviours that are atypical of the person and may also be inappropriate within their culture

mental wellbeing A state of mind characterised by emotional wellbeing, a capacity to establish and maintain constructive relationships, and high levels of functioning enabling a person to think clearly and logically to cope with the ordinary demands and stresses of life

mindfulness meditation A meditation practice in which a person focuses attention on their breathing, with thoughts, feelings and sensations being experienced freely as they arise and without judgement; intended to enable people to become highly attentive to sensory information and to focus on each moment as it occurs

negative reinforcement Occurs when the likelihood of a response recurring increases because it is followed by an end to discomfort, or by the removal of an unpleasant event

normality A pattern of thoughts, feelings or behaviour that conforms to what is considered to be acceptable or what can be expected to happen in most circumstances. In a clinical context, it is a condition that does not require treatment or assistance

operant conditioning A learning process in which the likelihood of a behaviour being repeated is determined by the consequences of that behaviour

perpetuating risk factors Longer-term, ongoing factors that tend to maintain the psychological problem after it has been elicited, inhibiting recovery by prolonging the disorder and preventing its resolution

precipitating risk factors The current specific personal or situational dynamics that occur shortly before the development of a disorder and appear to be the triggers for the onset or exacerbation of the psychological problems or behavioural response

predisposing risk factors Those aspects within an individual's characteristics (many of them operating from a person's formative early years) that increase their vulnerability to other influences that act close to the time of the development and presentation of psychological problems

prognosis A prediction about the probable course and outcome of a disorder

protective factors An individual's strengths, resilience, social supports and positive patterns of behaviour that prevent or reduce the likelihood or severity of conditions due to the risk factors present

psychiatrist A medical doctor who has received additional training in the study and treatment of mental illness and emotional disorders

psychoeducation A psychosocial approach in which a person experiencing a mental health problem or disorder and their family are provided with information to help them understand the condition and how they can contribute to managing it

psychological factors Within the biopsychosocial framework, the cognitive and behavioural influences that can have an impact on an individual's wellbeing (e.g. memory, thinking and reasoning skills, emotions)

psychopathology The scientific study of the causes and treatments of mental disorders or any psychological syndrome that impairs an individual's ability to adapt and function effectively in a variety of conditions

resilience A person's ability to successfully adapt to stress and cope with adversity, influenced by coping strategies, adaptive ways of thinking and social connectedness

self-determination Each person's ability to make choices and decisions that enable them to control their own life and manage their own psychological health and wellbeing

sleep A regular, naturally occurring altered state of consciousness required for restorative functions that enable physical and mental wellbeing

social and emotional wellbeing (SEWB) Preferred by many Aboriginal and Torres Strait Islander peoples, this term provides a multidimensional, holistic understanding of life and health, both physical and mental, within a framework that places Indigenous perspectives, connections and culture at the centre

social stigma A negative social attitude about a characteristic of a person or social group that implies some form of deficiency, often leading to unfair discrimination against or exclusion of the person or social group

social support The assistance and comfort we receive from people in our social network when we are facing a stressful or challenging situation, from family members and friends through to support groups and social institutions

social wellbeing The ability to form and maintain effective relationships, work productively and cooperatively with others, and be able to contribute to the community

sociocultural factors Within the biopsychosocial framework, the influences within an individual's social network that can have an impact on their wellbeing (e.g. relationships, family supports, life events)

specific environmental triggers A specific object, event and/or experience that can act as a stimulus for the onset of a particular condition

specific phobia An anxiety disorder characterised by an intense, excessive, irrational or persistent fear directed towards a particular object, situation or event, which causes significant distress or interferes with everyday functioning

stress An unpleasant state of physiological and/or psychological tension that is experienced when a situation is perceived as threatening to an individual's wellbeing, and therefore may tax or exceed their ability to cope

stress response The physiological and/or psychological changes experienced when an individual is confronted by a stressor

systematic desensitisation A type of behaviour therapy that uses counterconditioning to reduce the anxiety a person experiences when in the presence of, or thinking about, a feared stimulus. It involves first learning specific muscle relaxation techniques and then practising these while the psychologist exposes the client to experiences with the feared stimulus by systematically increasing the intensity of the experience, beginning in the imagination and ending in reality. Counterconditioning refers to learning the relaxation response that is incompatible with the fear response

typical Behaviour that is usually displayed by an individual the majority of the time and/or is displayed by the majority of people

Revision summary

Use the following summary of syllabus dot points and key knowledge within Unit 4 Area of Study 2 to ensure that you have reviewed the content thoroughly. Provide a brief definition or comment for each item to demonstrate your understanding or code them using the traffic light system: green (all good), amber (needs some review) or red (priority area to review). Alternatively, write a follow-up strategy.

What influences mental wellbeing?	
Defining mental wellbeing	
• Definition of mental wellbeing	
• The biopsychosocial model of mental wellbeing and mental disorder	
• Levels of functioning	
• Resilience	
• Coping with and managing change and uncertainty	
• Social and emotional wellbeing (SEWB) as a multidimensional and holistic framework for Aboriginal and Torres Strait Islander peoples	
• Mental wellbeing as a continuum	
• Variations for individuals experiencing stress, anxiety and phobia	
Application of a biopsychosocial approach to explain specific phobias	
• The relative influences of contributing factors to the development of specific phobia	
• Biological contributing factors	
– GABA (gamma-aminobutyric acid) dysfunction	

– Long-term potentiation	
• Psychological contributing factors	
– Behavioural models	
▪ Precipitation by classical conditioning	
▪ Perpetuation by operant conditioning	
– Cognitive biases	
▪ Memory bias	
▪ Catastrophic thinking	
• Social contributing factors	
– Specific environmental triggers	
– Stigma around seeking treatment	

• Evidence-based interventions and their use for specific phobia	
• Biological interventions	
– The use of short-acting anti-anxiety benzodiazepine agents (GABA agonists) in the management of phobic anxiety	
• Breathing retraining	
• Psychological interventions	
– Psychotherapeutic treatments	
– Cognitive behavioural therapy (CBT)	
– Systematic desensitisation	
• Social interventions	
– Psychoeducation for families/supporters	
– Challenging unrealistic or anxious thoughts	
• Discouraging avoidance behaviours	

Maintenance of mental wellbeing	
• The role of protective factors in maintaining mental wellbeing	
– Adequate nutritional intake	
– Adequate hydration	
– Adequate sleep	
– Cognitive behavioural strategies	
– Mindfulness meditation	
– Authentic and energising support from family, friends and community	
• Cultural determinants as integral for the maintenance of wellbeing in Aboriginal and Torres Strait Islander peoples	

Exam practice

Defining mental wellbeing

Answers start on page 251.

Multiple-choice questions

Question 1

The biopsychosocial framework seeks to describe and explain how biological, psychological and social factors interact to influence a person's

A   biological, physical and social factors interact to influence a person's physical and mental health.

B   biological, psychological and social factors interact to influence a person's psychological and mental health.

C   biological, physical and social factors interact to influence a person's psychological and mental health.

D   biological, psychological and social factors interact to influence a person's physical and mental health.

Question 2

If a GP was attempting to diagnose a mental disorder without considering the biopsychosocial model, they would be least likely to notice

A   severe emotional disturbances.

B   an unhealthy family environment.

C   a degenerative brain disorder.

D   hallucinations.

Question 3 ©VCAA 2018 SA Q34

What are the typical characteristics of a mentally healthy person in terms of their levels of functioning and social and emotional wellbeing?

	Functioning	Social wellbeing	Emotional wellbeing
A	Successfully accomplishes tasks	Shows respect for other people	Manages stress reactions
B	Manages stress reactions	Spends time with family	Works independently and with others
C	Overcomes problems	Demonstrates self-confidence when alone or with others	Respects the cultural identity of others
D	Demonstrates self-confidence when alone or with others	Undertakes everyday social interactions	Displays a positive attitude

Use the following information to answer Questions 4 and 5.

Glen lost his job just after his third child was born. Glen felt overwhelmed by the demands of being a parent while also being unemployed. The local council had also closed the park closest to his house. Glen really missed the opportunities that the park had provided for his older children to play and for him to spend time with other fathers.

Despite these challenges, Glen had strong support from his family and friends, and he was able to enjoy daily events related to being a father. He also actively looked for new employment opportunities and organised a surprise party for his own father's 70th birthday.

Question 4 ©VCAA 2017 SA Q39

Glen would be considered mentally healthy because he

A avoided stressful situations.

B was unable to focus on the needs of his family.

C received social and psychological support from his family and friends.

D worked towards goals in the face of stressors and disappointments in his life.

Question 5 ©VCAA 2017 SA Q42

To help Glen manage his feelings of being overwhelmed by fatherhood, Glen's doctor suggested that he use a combined biopsychosocial approach.

Which of the following includes all aspects of the biopsychosocial approach?

A Improving sleep strategies, challenging unrealistic thoughts and accessing social support

B Reducing stigma, and improving sleep strategies and personal relationships

C Challenging unrealistic thoughts, strategies and advice provided by family

D Improving personal relationships and self-efficacy, and reducing stigma

Question 6

Kumi was very upset when her dog died after a long illness. Despite this occurring months ago, she is still distressed, refuses to go out with her friends and cries herself to sleep every night. Kumi's behaviour would best be described as indicating a mental health problem (or worse) because of

A interference with normal functioning.

B being out of contact with reality.

C deviance from social norms.

D the presence of an underlying physiological condition.

Question 7

Psychologists use a number of criteria to help them determine whether someone is suffering from a disorder. These include a combination of all of the following **except**

A the presence of distress.

B impaired functioning.

C brain damage.

D atypical responses.

Question 8

What is the key criterion for identifying a person as having a mental disorder?

A The person has problems.

B The person's ideas challenge the status quo.

C The person's low level of functioning is below what is expected.

D The person makes other people feel uncomfortable.

Question 9 ©VCAA 2020 SA Q40

Which of the following accurately describes the impact on mental health of both external factors and resilience?

	External factors	Resilience
A	The effect depends on the number and the nature of the external factors.	Growth can occur even though setbacks occur.
B	External factors will not have an impact on the development of a mental health problem.	Resilience allows an individual to 'bounce back'.
C	Internal factors are likely to have a bigger impact than external factors.	The number of setbacks is likely to determine the level of resilience.
D	Social factors are likely to have a greater impact than emotional factors.	External factors do not influence resilience.

Question 10 ©VCAA 2020 SA Q44

Theodore lost his job 2 years before he intended to retire and it had a negative impact on his mood and ability to cope. He did not pay two electricity bills despite having sufficient funds. He became withdrawn while at his golf club and soon stopped playing. When he also started complaining of sleeping problems, his daughter encouraged him to see his family doctor with her.

To try to improve Theodore's resilience, the doctor decided to focus on biological, psychological and social factors. Which of the following is a combination of the most likely strategies the doctor would initially encourage or use?

	Biological	Psychological	Social
A	Prescribing medication	Challenging Theodore's negative thought patterns	Reminding Theodore about his supportive family
B	Getting Theodore's friends to bring him meals	Getting Theodore to join the seniors' club	Getting Theodore to start an exercise program
C	Organising genetic testing	Getting Theodore to make lists of the things he needs to do	Taking Theodore to a nutritionist
D	Organising nutritious meals	Challenging Theodore's negative thinking about the future	Getting Theodore's friends to visit him regularly

Question 11

The concept of social and emotional wellbeing (SEWB)

A is only concerned with the mental health of the individual.

B only focuses on these dimensions and ignores biological and cognitive factors.

C provides a totally different approach to health in comparison to the biopsychosocial model.

D acknowledges that factors from several domains may combine to influence mental wellbeing.

Question 12

Benita is experiencing dizziness, muscle tightness, shaking and tremors. For no apparent reason, she is feeling apprehensive that something is going to go wrong in the future. These symptoms most resemble those found in cases of

A Alzheimer's disease.

B anxiety disorders.

C Parkinson's disease.

D stress.

Question 13 ©VCAA 2018 SA Q37

Anxiety can be distinguished from phobia because only anxiety

A involves distress.
B can be helpful in mild amounts.
C triggers the fight–flight–freeze response.
D is influenced by biological, psychological and social factors.

Question 14

Shae has been experiencing ongoing anxiety attacks for some time. She describes the symptoms to her GP, who suspects that Shae may have a neurochemical basis resulting from insufficient levels of

A GABA.
B dopamine.
C noradrenaline.
D glutamate.

Question 15

A phobia is best defined as

A a rational fear of a specific object or situation.
B an irrational fear of a specific object or situation.
C a response that is developed through the repeated pairing of fear and illness.
D a response that automatically occurs when an unconditioned stimulus is presented.

Short-answer questions

Question 16 (16 marks)

a Describe what is meant by 'mental health'. — 4 marks
b Outline the characteristics of 'mental health problems'. — 3 marks
c Explain the meaning of 'mental illness'. — 4 marks
d Define 'psychological resilience'. — 2 marks
e Outline what is meant by 'social and emotional wellbeing' for Aboriginal and Torres Strait Islander peoples. — 3 marks

Question 17 (5 marks)

a List three criteria that could be used to determine whether a person might have a psychological disorder. — 3 marks
b Describe some of the general problems in classifying a behaviour as a mental disorder. — 2 marks

Question 18 (3 marks) ©VCAA 2015 SB Q5

One year ago, Toby's wife died. For the past 6 months, Toby has been acting out of character. Despite being physically in good health, he has not left the house for several weeks and has asked his parents to shop for groceries for him. Whenever his parents visit, they notice that he has not showered for days and is often wearing the same clothes. He lost his job because of extended absences and his friends are concerned as Toby is no longer responding to their text messages or telephone calls. He has also shown a lack of interest in physical activity despite previously completing many marathons.

Give three reasons why these behaviours may cause Toby's psychologist to conclude that he has a mental illness.

Question 19 (8 marks)

For each of the following scenarios, identify which letter best indicates where each individual would be placed on the mental health continuum and explain why.

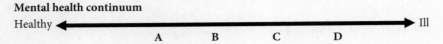

a Eliana has been experiencing a consistently low mood. For the past few months, she has not engaged in social activities and has not attended school for her VCE classes. 2 marks

b Chen is entering his final year of university and needs a driver's licence to travel to work placements as a part of his course. By focusing on the need to pass, he puts so much pressure on himself that he keeps failing his driving test. Despite the extra costs involved, which are putting a strain on his finances, he continues to take driving lessons and books another test. 2 marks

c Stefan describes things as generally going well at the moment. He is coping with the workload of his Year 12 studies and is able to balance this with an active social life with his girlfriend. He is, however, nervous about having to play a piano solo at the school's musical recital next week. 2 marks

d Voula is finding that she is struggling to manage both her university studies and working part-time to cover her costs, so she is considering whether she should defer her course. As a result, she has had trouble sleeping for the past few weeks. 2 marks

Question 20 (3 marks)

Provide **three** characteristics that differentiate phobias from stress and anxiety.

Application of a biopsychosocial approach to explain specific phobia

Answers start on page 254.

Multiple-choice questions

Question 1

The difference between a phobia and a normal fear is that a phobia

A can trigger a maladaptive response.

B is characterised by an increase in heart rate.

C only takes place when the actual object is present.

D involves a cognitive component in addition to behavioural responses.

Question 2 ©VCAA 2017 SA Q15

Sophia has developed a specific phobia of dogs and cries whenever she sees one.

With reference to gamma-aminobutyric acid (GABA), it is most likely that Sophia has

A an excess of this excitatory neurotransmitter.

B an excess of this inhibitory neurotransmitter.

C a deficiency of this excitatory neurotransmitter.

D a deficiency of this inhibitory neurotransmitter.

Question 3

Mae was on her way to visit a friend in his high-rise apartment when there was a blackout while she was in the elevator, leaving her trapped for several hours until the power went back on. Since that time, Mae has had a terrible fear of enclosed spaces.

Identify which of the following statements best describes the influence of long-term potentiation as a contributing factor to the development of Mae's phobia.

A The episodic memories of being trapped in the elevator have been transferred into her long-term memory.

B Connections within the neural pathways representing the association between enclosed spaces and Mae's fear of them have been strengthened.

C The neural impulses that fired when Mae was trapped in the elevator are still displaying long-term activation.

D The recollection of her experience in the elevator has the potential to cause distress in the long term.

Question 4

Behavioural models explain phobias as precipitated through _____ of a fear response and perpetuated by _____ of avoidance (escape) behaviour.

A positive reinforcement; negative reinforcement

B observational learning; positive reinforcement

C negative reinforcement; positive reinforcement

D classical conditioning; operant conditioning

Question 5

According to classical conditioning, arachnophobia (a strong fear of spiders), develops through

A the association of an unconditioned response (fear) with an unconditioned stimulus (spiders).

B reinforcement when the strong fear that occurs with spiders is decreased when the person avoids any contact with them.

C the association of a conditioned stimulus (spiders) with an experience that caused intense fear.

D an innate reflexive response of strong fear whenever the person comes into contact with spiders.

Question 6

Djamila becomes anxious whenever she walks by a particular alley. Two years ago, a man held a gun to her head and stole her purse in this alley. In this example, the conditioned response is

A the alley.

B feeling anxious when she walks by the alley.

C being robbed by a man with a gun.

D feeling anxious when she was robbed by a man with a gun.

Question 7

Ever since watching the movie *Deep Blue Sea*, Kaito has developed an intense and irrational fear of sharks, despite never having encountered one in real life. He is afraid to go swimming in the sea because he thinks sharks will attack him and that he will get bitten by one. He experiences fear even when he sees a picture of a shark in a book or on television.

Kaito's thoughts about being the victim of a shark attack is an example of

A operant conditioning.

B memory bias.

C catastrophic thinking.

D an environmental trigger.

Question 8 ©VCAA 2019 SA Q40

Which one of the following does **not** explain why stigma is viewed as a social risk factor in the development and progression of a mental illness?

A People with a mental illness avoid seeking support.

B People with a mental illness perceive themselves as different.

C People with a mental illness perceive having a mental illness as shameful.

D Stigma supports negative stereotypes about mental illness in the community.

Question 9

Benzodiazepines are used to treat phobic patients by

A activating postsynaptic neurons in the brain to reduce arousal and calm the body.

B inhibiting presynaptic neurons in the brain to activate the body and increase arousal.

C activating presynaptic neurons in the brain to increase arousal and activate the body.

D inhibiting postsynaptic neurons in the brain to calm the body and reduce arousal.

Question 10

Breathing retraining is an example of a

A biological approach to reduce physiological arousal.

B psychological approach because it reduces psychological arousal by getting the individual to focus on something else.

C social approach because it involves a distraction from the stressful situation while being coached by another person during the training.

D psychological approach because it uses mind over matter to reduce physiological arousal.

Question 11

Exposure therapy for phobias is focused on extinguishing the

A operantly conditioned avoidance response.

B classically conditioned avoidance response.

C operantly conditioned fear response.

D classically conditioned fear response.

Question 12

In systematic desensitisation to treat phobias, the fear that has become associated with a specific object or event is reduced through

A negative reinforcement.

B delivery of an aversive consequence.

C counterconditioning and extinction.

D operant conditioning.

Question 13

In systematic desensitisation

A the person with the phobia is challenged in terms of maladaptive cognitions.

B phobic stimuli are arranged in a hierarchy from least- to most-feared items.

C the person with the phobia is immediately brought into direct contact with their most feared object or situation and kept in contact with it until their fear and associated anxiety dissipates.

D All of the above.

Question 14

Kaia is a psychologist who works with her clients to help them challenge their irrational thoughts and practise more adaptive behaviours to replace their old, problematic ways of dealing with things.

This type of therapeutic approach employed by Kaia is

A psychoeducation.
B systematic desensitisation.
C mindfulness techniques.
D cognitive behavioural therapy.

Short-answer questions

Question 15 (2 marks)

a Define 'phobia'. — 1 mark
b Explain why diagnostic systems classify a phobia as being a type of anxiety disorder. — 1 mark

Question 16 (7 marks)

a Outline the influence of gamma-aminobutyric acid (GABA) as a contributing factor in the development of specific phobia. — 2 marks
b Describe how benzodiazepine agents might be used as part of a treatment plan for phobic anxiety. — 3 marks
c Summarise the benefits of breathing retraining for someone with a specific phobia. — 2 marks

Question 17 (12 marks)

a Using appropriate terminology, clarify how behaviourist theories would describe the conditioning of phobias. — 4 marks
b Explain the principles of learning on which systematic desensitisation is based. — 3 marks
c Briefly outline how systematic desensitisation is used as a psychotherapeutic treatment of phobia. — 2 marks
d Describe the steps involved in systematic desensitisation. — 3 marks

Question 18 (7 marks)

a Clarify how the process of long-term potentiation is a contributing factor in the development and maintenance of specific phobia. — 2 marks
b Explain how cognitive bias, including memory bias and catastrophic thinking, contributes to phobic disorders. — 3 marks
c Outline how cognitive behavioural therapy (CBT) can be applied to help reduce anxiety for people with phobias. — 2 marks

Question 19 (9 marks)

a Describe how specific environmental stimuli can trigger a phobia. Support your answer with an appropriate example. — 2 marks
b Using appropriate terminology, explain how some phobias may be transmitted within families. — 2 marks
c Outline why social stigma could be a potential issue for those with phobias. — 2 marks
d Summarise how the process of psychoeducation would be applied within a phobic context. — 3 marks

Maintenance of mental wellbeing

Answers start on page 258.

Multiple-choice questions

Question 1

Exercise, good nutrition and an adequate amount of sleep are _____ against the development and progression of mental health disorders.

A predisposing risk factors

B precipitating risk factors

C perpetuating risk factors

D protective factors

Question 2

Exercise, meditation and relaxation are important factors in stress management because they

A reinforce the fight–flight–freeze response.

B distract attention from stressful situations.

C can alleviate the physiological arousal associated with stress.

D remove the actual source of stress.

Question 3

An example of a psychological protective factor for mental wellbeing is

A challenging negative or unrealistic thinking.

B active participation in cultural activities.

C support from family, friends and community.

D eating a balanced diet to provide essential nutrients for the brain to function effectively.

Question 4

Mindfulness meditation

A focuses an individual's concentration away from thoughts and feelings by repeating a word or mantra over and over in order to generate a sense of relaxation.

B reduces psychological arousal by getting the individual to focus on something in the current moment.

C causes an increase in slow, delta brainwave activity which lasts throughout the day.

D is a psychological approach because it uses mind over matter to reduce physiological arousal.

Short-answer questions

Question 5 (7 marks)

a Outline the role protective factors play in relation to an individual's mental health.   1 mark

b Summarise how each of the following aspects of a healthy lifestyle are factors that contribute to an individual's mental wellbeing.

    i Nutrition   1 mark

    ii Hydration   1 mark

    iii Sleep   1 mark

    iv Exercise   1 mark

c Outline how an unhealthy lifestyle can affect an individual's physical and mental wellbeing.   2 marks

Question 6 (3 marks)

a Summarise the general aim of cognitive behavioural strategies. 2 marks

b Explain how the process of mindfulness meditation can help clarify our thinking and reduce stress to maintain mental and physical wellbeing. 1 mark

Question 7 (4 marks)

Describe two examples of activities that develop and maintain cultural continuity, and how each can contribute to the maintenance of social and emotional wellbeing for Aboriginal and Torres Strait Islander peoples.

Question 8 (9 marks) ©VCAA 2017 SB Q6

Zac lives with his parents and contributes financially to the household by paying all of the rent. Zac was quite stressed about the expectations placed on him to provide financially for the family. Recently, his grandfather, who lives with them, became quite ill. Zac soon felt unable to cope with these stressors and tried to manage his distress through substance use. He subsequently lost his job and isolated himself from friends and family. Finally, his sister encouraged him to seek professional support from a psychologist, who then diagnosed him with a mental health disorder.

a Identify one biological risk factor and one psychological risk factor that are present in the scenario, and outline how each of these factors may have contributed to the development and/or progression of Zac's mental health disorder. 4 marks

b Identify a different type of coping strategy that Zac could have used when he lost his job. 1 mark

c Describe **one** social strategy that Zac's psychologist could recommend to Zac to increase his resilience. 2 marks

d Identify **one** possible source of stigma and how this stigma could be a barrier to Zac accessing treatment for his mental health disorder. 2 marks

Chapter 5
Area of Study 3: How is scientific inquiry used to investigate mental processes and psychological functioning?

Area of Study summary

Within this area of study, you will need to apply the knowledge and skills you have learned in VCE Psychology about research methods by designing and implementing a practical investigation related to mental processes and psychological functioning. This research involves the collection of primary data relating to the ideas and skills developed in Unit 3, Unit 4 or across Units 3 and 4.

The investigation requires you to:

- develop a question
- articulate an aim
- formulate a research hypothesis
- plan a course of action that attempts to answer the question
- take into account safety and ethical considerations
- undertake an investigation either in the laboratory and/or in the field
- collect primary quantitative data
- analyse and evaluate the data
- identify any limitations within the data and methodology
- relate your results to relevant concepts within the course
- reach a conclusion that addresses your question
- suggest further investigations which may be performed to explore the concept.

Your investigation will be presented as a structured scientific poster according to a template specified by the Victorian Curriculum and Assessment Authority. You are also required to maintain a logbook documenting all the practical work you did at each step within your research. This record is essential as part of the assessment process and to authenticate that the work produced is in fact your own.

Area of Study 3 Outcome 3

On completion of this outcome, you should be able to:

- design and conduct a scientific investigation related to mental processes and psychological functioning
- present an aim, methodology and method, results, discussion and conclusion in a scientific poster.

The key science skills demonstrated in this outcome are:

- develop aims and questions, formulate hypotheses and make predictions
- plan and conduct investigations
- comply with safety and ethical guidelines
- generate, collate and record data
- analyse and evaluate data and investigation methods.

Adapted from *VCE Psychology Study Design (2023–2027)* © copyright 2022, Victorian Curriculum and Assessment Authority

As a part of your VCE Psychology course, you will be required to undertake a scientific investigation related to mental processes and psychological functioning drawing on knowledge and related key science skills developed in Unit 3, Unit 4 or across Units 3 and 4.

Like any of the other learning outcomes, this area of study is examinable and will be covered in questions on the end-of-year examination.

The first part of this chapter will summarise key knowledge and definitions pertaining to research methods in psychology that will help you prepare for the investigation and the examination. The rest of this chapter will provide some guide notes for the investigation and the research poster.

5.1 The research process

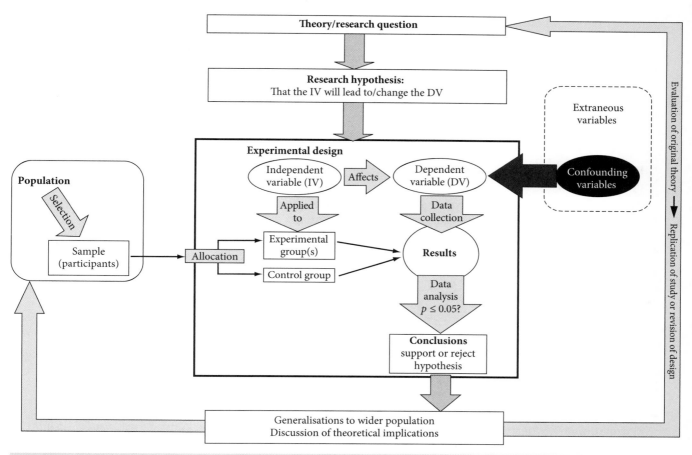

FIGURE 5.1 The research process

A **theory** is a tentative explanation for a phenomenon and attempts to describe how ideas fit together. In order to investigate a theory, a researcher must formulate a **hypothesis**; that is, a tentative, testable assumption or reasoned prediction as to how the independent variable(s) or treatment(s) will affect the dependent variable(s).

A **variable** is a factor that has the capacity to increase or decrease in amount or kind. Characteristics or factors that are fixed or unchanging cannot be considered to be variables in an experiment.

An **independent variable (IV)** refers to the treatment variable in an **experiment** that is deliberately varied or systematically manipulated by the experimenter in order to measure whether it produces a change in the **dependent variable (DV)**, as observed or measured by a change in the behaviour or performance of the **participants**.

If a theory or hypothesis proposes a cause-and-effect relationship, then the independent variable is perceived as the *cause*. Changes in the dependent variable are therefore perceived as being an *effect* of the manipulation of the independent variable.

5.1.1 Operational hypotheses

Because many of the variables in psychological research are abstract concepts, researchers must design their experiments so that the variables can be measured through observable behaviours.

An **operational hypothesis** expresses the experimental/research hypothesis in terms of how the researcher will determine the presence and levels of the variables under investigation; that is, how the experimenter is going to put his or her hypothesis into operation.

The operational hypothesis describes how variables will be observed, manipulated and measured by the experimenter, along with details of the **population** sampled.

> **Note**
> The concept of an operational hypothesis combines two dot points together – one to do with hypotheses and the other to do with operationalising variables.
>
> When asked to write an operational hypothesis, remember **IPOD** – as your statement should include the **I**ndependent variable, identification of the **P**articipants/**P**opulation of interest, how the variables will be **O**perationalised (put into practice within the research design) and the **D**ependent variable.

5.1.2 Methodologies used in psychological research

TABLE 5.1 Descriptions of the common methods used in psychological research

Research method	Brief description	Advantages	Disadvantages
Controlled experiment	The researcher systematically manipulates or changes a particular variable (IV) under controlled conditions while observing the resulting changes in another variable (DV).	• Identifies cause-and-effect relationships • Distinguishes between placebo effects and actual effects of a treatment or drug	• Can be artificial (lacking ecological validity), so results may not generalise to real-world situations
Case study	An in-depth study of an individual, a small group or an event, collecting data through interviews, **direct observation**, psychological testing or examination of documents and records to try to interpret the person's behaviour.	• Provides a good way to generate hypotheses • Yields data that other methods can't provide	• Information may be incomplete • Can be subjective and thus may yield biased results • Doesn't allow conclusions about cause-and-effect relationships
Correlational study	Research exploring the association between two or more variables. Statistical measures enable researchers to establish the strength and direction of the relationship between the variables, events or measures.	• Can identify which factors are more significant • Can be useful for generating hypotheses or directions for further research	• Doesn't allow for scientific control of variables to draw conclusions about cause-and-effect relationships • Experimenters cannot control extraneous variables • Doesn't explain the reason behind a relationship between two variables • The association may be coincidental or connected to another (possibly unidentified) variable

TABLE 5.1 cont.

Research method	Brief description	Advantages	Disadvantages
Naturalistic observation (fieldwork)	Researchers collect information by observing subjects unobtrusively within their natural environment, without interfering with or manipulating them in any way. Researchers create a record of events and note relationships among those events.	• Can be useful for generating hypotheses or directions for further research • Provides information about behaviour in the natural environment	• May be difficult to do unobtrusively – subjects may be aware of the observer and may act differently as a result • Doesn't allow for scientific control of variables to draw conclusions about cause-and-effect relationships • Experimenters cannot control extraneous variables • Can be time consuming and expensive • Open to the effects of observer bias
Laboratory observation	Allows researchers some degree of control over the environment and enables them to use sophisticated equipment to measure and record subjects' behaviour. They can use one-way mirrors or hidden recording devices to observe subjects more freely while remaining hidden themselves.	• Enables use of sophisticated equipment for measuring and recording behaviour • Can be useful for generating hypotheses	• Sometimes yields biased results • Carries the risk that observed behaviour is different from natural behaviour (i.e. lack external validity) • Doesn't allow conclusions about cause-and-effect relationships
Self-report	Through questioning or diaries, the researcher gets participants to reflect on and describe their experiences. Useful to explore the thought processes and motivations behind various feelings and behaviours.	• Yields data about behaviour that can't be observed directly	• Can be inaccurate as these rely on the honesty and/or the ability of individuals to verbalise the information • Can be very subjective and open to bias • Don't allow conclusions about cause-and-effect relationships
Questionnaires	Researchers give people a set of questions to answer in order to obtain information about a specific type of behaviour, experience or event.	• Can yield a lot of information • Provide a good way to generate hypotheses • Generally cheap and easy to do • Information can be gained without the need for direct contact, through mail, Internet etc.	
Interviews	Researchers ask people a set of questions in order to obtain information about a specific type of behaviour, experience or event.	• Can yield data about behaviour that can't be observed directly • Information can be clarified by asking the person to elaborate • Additional non-verbal information can be observed	
Rating scales	A set of categories designed to elicit quantitative or qualitative measures of a subjective attribute.	• Subjective data can be gathered in a quantitative measure to allow for statistical analysis	

TABLE 5.1 cont.

Research method	Brief description	Advantages	Disadvantages
Literature review Archival research	Collation and analysis of secondary data related to other researchers' scientific findings and/or viewpoints. This can involve going through records and/or journal articles detailing previous research, or can entail a variation of the case study method, whereby the researcher goes through historical records, such as patient files, to gather data for analysis.	• Can yield a lot of information • Can be useful for generating hypotheses or directions for further research • Can provide background information to help answer a question or explain observed events • Yields data that other methods can't provide, especially in experiments or case studies that can't be replicated for ethical and/or practical reasons • Can provide data regarding generational differences of changes over time without having to undertake a **longitudinal study** • Can be useful for generating hypotheses or directions for further research	• Information may be incomplete • Data could be subjective, especially in patient files, and thus may yield biased results • Doesn't allow conclusions about cause-and-effect relationships
Computer simulation	Researchers use computer programs to mimic behaviour, allowing them to manipulate variables without fear of harming the subject under study.	• Enables **modelling** and manipulation of variables that cannot be controlled because of the complexity, size, timing or accessibility of the phenomenon being studied • Enables exploration of behaviours in a way that does not expose the participant to physical threat or harm • Enables the ethical exploration of animal behaviour without the need for actual live subjects	• Can be artificial (lacking ecological validity), so results may not generalise to real-world situations • Depends on the reliability and breadth of the information incorporated into the program • May not be able to reproduce all of the relevant variables that may affect the behaviour under study

5.1.3 Ways to minimise the effects of extraneous variables

Extraneous variables are those elements outside of the experimental design that have the potential to influence the dependent variable and so could have an unwanted or unintentional effect on the results of the study. Such variables may be associated with characteristics of the participants, such as age, gender and ethnicity, or the researcher's behaviour, or the experimental design itself, especially if non-standardised instructions are given.

As such, extraneous variables are conditions that a researcher wants to prevent from influencing the outcome of the experiment.

The two types of extraneous variables are controlled and uncontrolled variables.

Controlled variables are those extraneous variables whose influence has been removed (or at least diminished) from the research via sampling (to control for participant characteristics), procedure (by using standardised instructions and procedures to keep it constant) or by the use of statistical methods of control.

Uncontrolled variables are variables that have influenced the result because their presence was not accounted for (and removed) in the experimental method. Uncontrolled variables that do cause a systematic variation (change) in the value of the dependent variable are termed **confounding variables**. Such variables interfere with the **validity** of the experiment by providing alternative explanations for the results.

For example, weather is external to an experiment and not usually a factor that needs to be considered. But consider the following scenario. Early in Term 1, two classes from the same cohort undergo an experiment involving performance on several different tasks. Temperatures over the 4 days leading up to the activity were all over 40°C and the school's air conditioning had broken down. The first group performed the experiment in the last period of the day in a room on the top floor of the building that directly faced the sun. The second group did the activity the next morning, after the cool change had come through the night before, and were in a room on the lowest floor which, being partially below ground level, had remained at an even, comfortable temperature. Results were collated for the two groups and showed significant differences in their performance. In this case, the differences were not due to participant characteristics, but due to the effects of environmental factors at the time of testing. As such, the weather (heat and humidity) and potentially the time of day were uncontrolled variables that had confounded the results. In order to eliminate these factors, the experimenter needed to make them controlled variables by doing the experiment under the same conditions in a cool room at the same time of day.

> **Note**
> Extraneous variables *may* affect the DV.
> Confounding variables are a subset of extraneous variables that *do* affect the results.

Experiments can demonstrate that the IV influences the DV and, if the research is carefully designed, we can generalise the results of laboratory experiments to phenomena in everyday life. However, as the research design may be flawed, the study should be repeated a number of times to clearly display a causal relationship between the IV and DV before conclusions can be applied to the general population.

5.1.4 Participant selection

The process of selecting participants for a study from the population of interest is known as **sampling**. In order to generalise the results back to the population, **samples** should be as representative (or typical) of the population as possible. They should include sufficient participants to portray the variety of individuals within the total group from which the sample is drawn.

Target population

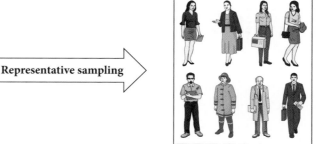

Representative sampling

Sample

Random sampling allocates participants from the population of interest to form part of the sample, such that each member of the population has an equal chance of being selected. If the sample is of sufficient size, it is usually representative of the population.

Stratified sampling attempts to prevent biases by making the sample more representative of the population. This involves identifying some of the relevant factors (strata) present in the population, such as age, sex or income level, and then selecting a separate sample from each stratum in the same proportions as the population. Stratified sampling can be very time consuming, as the relevant characteristics and/or factors need to be identified and their proportions in the target population calculated.

Convenience sampling is the process of selecting participants who are easily obtainable. For example, newspaper polls, radio station phone-ins, approaching individuals in a shopping centre or doing an activity within a class. Such a sample may not be representative of the population and may be biased because individuals actively volunteer to participate in the study or are drawn from a narrow stratum of the population.

Convenience sample

5.1.5 Participant allocation

While random sampling is a method used to *select* participants for an experiment, random allocation (or assignment) is a means of *experimental control* that is used to place participants into groups within the experiment. Random allocation ensures that the participants selected are equally likely to be placed in any of the groups in the experiment. By doing so, similar groups will be obtained before the independent variable is administrated so that its effect can be measured accurately.

In an **experimental group**, the participants are exposed to the independent variable in order to observe its effects on their behaviour or experience.

Most experiments also incorporate a **control group**, which is a group of participants within an experiment with similar characteristics to the experimental group, but who do not receive the independent variable. The control group is used as a standard for comparison against the experimental group. This enables the researcher to determine or conclude whether it was the independent variable that has affected the dependent variable.

5.1.6 Placebo effect

A **placebo** is a fake treatment, often used in medical research in the form of sugar tablets or saline injections. It acts as a control condition in experiments to counter the effect of participants knowing they have taken something.

In psychological research, the **placebo effect** may occur when a participant's response (the dependent variable) is influenced not by a specific procedure that has been administered to elicit that behaviour or physiological response (the independent variable), but rather by the expectancy of how they should behave. A placebo is therefore used as a controlled variable to limit the difference between certain expectations being experienced by members of the experimental group that are not experienced by the control group.

To control the potentially confounding influence of the placebo effect, a researcher could employ a **single-blind procedure**. In this procedure, the researcher would allocate participants to either the experimental or control group, but not tell them which condition they were in. This reduces the chances of participants guessing the aims of the study. **Experimenter bias**, however, can still be present.

5.1.7 Experimenter effect

An **experimenter effect** occurs when the unconscious expectations, personal characteristics or treatment of the **data** by the experimenter adversely affects the dependent variable, which biases the results. This can occur if the experimental and control groups are treated differently in ways other than those related to the independent variable.

Experimenter bias occurs when the researcher's unconscious expectations or motivation may influence, and therefore distort, their **observations** of data. To eliminate this, experimenters can use a **double-blind procedure** in which neither the experimenter nor the participants know which subjects have been allocated to the experimental or control group. A third person controls the placement of groups.

5.1.8 An overview of three experimental designs

Repeated-measures design

In the **repeated-measures design**, one group of participants undertakes *both* experimental conditions (i.e. the experimental and the control condition).

The advantage of a repeated-measures design is that subject variables are highly controlled, as they are kept constant between conditions. For example, the intelligence levels and personality traits are identical in both groups, as the same people are used. Another advantage is that fewer participants are required, as each is used more than once, making the research more economical.

The disadvantage of this design is that repetition effects can occur. The participants may be bored and/or fatigued when performing the second condition, as they have already performed a task.

A further disadvantage of this method is that the practice effect could make the results invalid. For example, if students were taught touch typing using two different methods, their performance using the second method may be superior because of practice using the first method. In situations such as these, the repeated-measures design is inappropriate.

The disadvantage of the effects of repetition may be overcome by *counterbalancing* the groups by placing half the participants in the experimental condition first and the other half in the control condition first, thereby balancing the order effects equally between both conditions.

Matched-participants design

The **matched-participants design** (or matched-pairs design) involves placing equivalent pairs of participants in each group. Participants in each condition are paired according to any important variables which, if not controlled, may have a confounding effect on their performance in the research. These include intelligence, experience, age and sex.

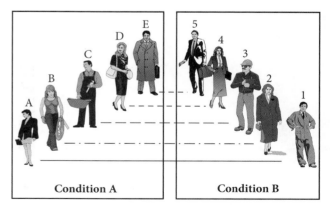

The matched-participants design eliminates order effects and demand characteristics (such as fatigue), as the participants only participate in one experimental condition. Although this design attempts to keep subject variables constant between conditions, participants can never be perfectly matched in every respect. The process of matching participants is also very time consuming and can be difficult to achieve.

Independent-groups design

The **independent-groups design** uses random allocation to groups as a means of controlling confounding variables. **Random allocation** (also termed random assignment) ensures that participants selected for the experiment are equally likely to be placed in any of the experimental or control groups.

Condition A

Condition B

The independent-groups design also eliminates order effects and demand characteristics as the participants only participate in one experimental condition, but subject variables could still occur despite random assignment. This method of research design is therefore the least effective in minimising the effects of extraneous variables and is only used when it is impossible to use the matched-participants design.

5.1.9 Statistics

Quantitative measures should take the form of numerical values and can be expressed in units of measurement (e.g. a test score or physiological measurements, such as height and weight).

> **Note**
> When we think of **quantity**, we think of amounts or numbers, which is the way **quantitative** data is presented.
> If the data has more **quality**, it provides more detail or description, which is how **qualitative** data is usually presented.

Qualitative measures are factual descriptions about the characteristics of the subjects' behaviour. Qualitative data should be in the form of words (e.g. case studies or responses to open-ended questions).

Subjective data is obtained by self-report measures in which subjects give verbal or written responses to a series of research questions (e.g. a survey on attitudes or the description of a dream).

Objective data is data that has been gathered using systematic observation that is not influenced by any personal bias (e.g. brain images or a test score).

Descriptive statistics help to organise data in order to provide a summary that can be easily interpreted or communicated. Descriptive statistics provide information to the researchers about their research sample that may affect their choice of **inferential statistics**.

A graph is preferable to a table of raw data, but the type of graph you use must be appropriate. The following examples indicate the elements that should be included as well as advice on when they should be used.

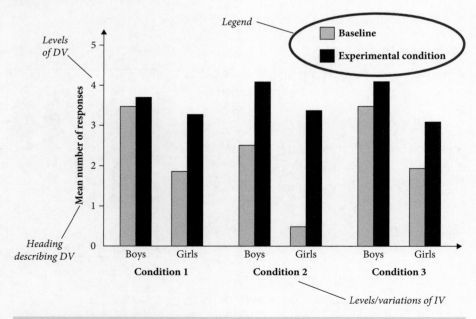

FIGURE 5.2 An example of a bar graph

Column and bar graphs (Figure 5.2) are useful for comparing several distinct data sets. Such graphs can show data for all levels or conditions of the IV, even those with zero values. The column width is equal so that the area is proportional to the amount of data in each level.

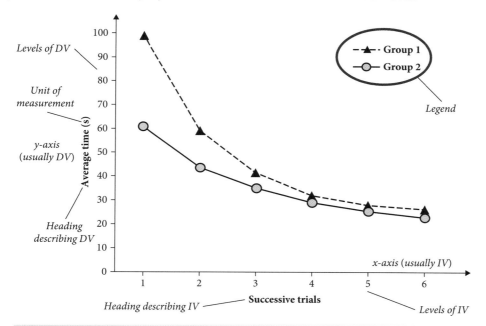

FIGURE 5.3 An example of a line graph

Line graphs (also known as frequency polygons) (Figure 5.3) are used to portray changes over time for continuous data. These graphs allow two or more sets of data to be shown and compared on the same graph. These graphs should not be used for categorical data.

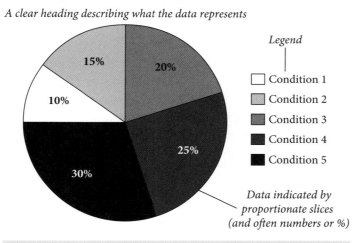

FIGURE 5.4 An example of a pie chart

Pie charts (Figure 5.4) are used to represent the relative proportions (percentages) of different characteristics within a whole set of data or group or population.

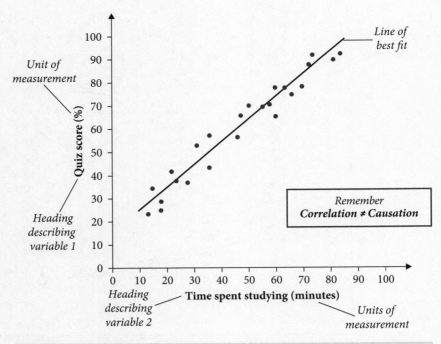

FIGURE 5.5 An example of a scatter plot

A scatter plot (Figure 5.5) is used to show a relationship (correlation) between two variables. The example represents a strong positive correlation – as one variable increases, so does the other. If the line of best fit were going down, it would be a negative correlation – when one variable is high, the other is low, and vice versa.

Measures of central tendency: mean, median and mode

Measures of central tendency are statistics that indicate information about the middle scores in a set of data. These middle scores are used as indicative measures for trends in the population and are most useful when data forms a relatively normal distribution. Measures of central tendency include the mean, median and mode.

The **mean** is the average score, which is obtained by adding up all the scores in the set of data and dividing this sum by the total number of scores present. For example, if the data set was 5, 6, 1, 7, 3, 8, 9, 6, 6 and 4, then the mean would be 55 ÷ 10 = 5.5.

The mean is the most sensitive measure of central tendency; however, it can be distorted by extreme values (**outliers**). For example, if we change one of the scores in the previous data set to be 5, 6, 1, 7, 3, 8, 9, 6, 6 and 100, then the mean would become 151 ÷ 10 = 15.1, which is not a meaningful measure of central tendency, as it is well above all but one of the scores. Such a set of scores would lead to a positively skewed distribution (see Figure 5.6).

The **median** is the middle point in a set of data when it has been put in order. For example, if the data set above was written in order, it would be 1, 3, 4, 5, 6, 6, 6, 7, 8 and 9, and the median would be 6. (Note: If the sample size is even, then the median would be the average of the middle two numbers.)

The **mode** is the most frequently occurring score in a set of data. For example, in the data set just given, the mode would be 6. Some distributions may have more than one mode.

The presence of extreme outliers can skew the distribution of scores, thereby affecting all of the measures of central tendency (as shown in the graphs in Figure 5.6). As such, they need to be identified and considered within any analysis and discussion of the data. (Note: For a negative skew, the outlier would be at the low end and the graph would be the mirror image of that shown in the examples.)

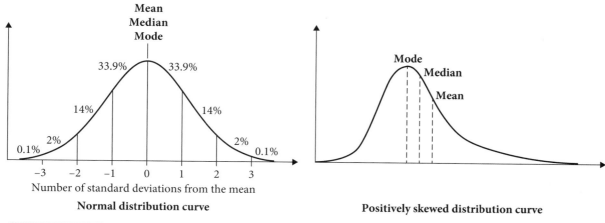

FIGURE 5.6 Examples of distribution curves

Measures of dispersion

The **range** indicates the dispersion, or difference, between the highest and lowest scores in a set of data. Although quick and easy to calculate, it can be distorted by extreme values.

The **standard deviation** is a measure of the variability of scores in a distribution indicating the average difference between the scores and their mean. The standard deviation is the most sensitive measure of dispersion, as it uses all the data available. The size of the standard deviation gives us an indication of the shape of the curve (a large standard deviation yields a flatter curve; small standard deviations mean a higher curve). A smaller standard deviation is indicative of more consistent results within the sample and suggests higher reliability of the data.

5.1.10 Reliability and validity of results

Researchers rely on the **accuracy** of results to show support or lack of support for their theory, and whether the findings of a study can be applied to the population of interest. If the research methods are faulty, the data collection and analysis will also be flawed.

Validity

If research is valid, then it truly explores and generates results that accurately measure the intended focus of the investigation. Experimental validity encompasses the entire empirical process and establishes whether the results obtained meet all the requirements of the scientific research method.

TABLE 5.2 Types of research validity

Type of validity	Description
Internal	A measure that ensures the experimental design is appropriate and effectively encompasses all the steps of the scientific research method, eliminating potential confounding variables in order to determine if a causal relationship exists between the independent variables and the dependent variables.
Predictive	A decision as to whether findings can be used to predict future behaviour based on assessing a group of subjects for a certain construct and then comparing them with results obtained at some point in the future.
Content	A qualitative measure of how well the substance of a test, including all its subtests and items, represents the construct it is designed to measure.
Construct	Whether an experiment actually measures what it is intended to measure to reflect the theory and **empirical evidence** on which it is based.
External	The extent that the conclusions drawn from the research can be generalised from a small sample group, often in laboratory surroundings, to truly represent and make predictions about other members from the population of interest in situations beyond the research study.

TABLE 5.2 cont.

Type of validity	Description
Population	Researchers ensure that the sample groups are as representative as possible, using random selection and large sample sizes to allow meaningful statistical analysis so that conclusions can be extrapolated to represent the population as a whole.
Ecological	A determination as to whether the research situation itself has influenced the behaviour being measured. The artificial nature of many studies in psychology can make generalisation difficult, as the experiment does not resemble the real-world situation.

Reliability

Reliability refers to the consistency, dependability and stability of the results obtained from a research study. Any significant results must be more than a one-off finding and should be inherently repeatable, yielding similar or compatible results under the same conditions in different experiments or research studies. This will reinforce the findings and ensure that the wider scientific community will accept the hypothesis.

TABLE 5.3 Types of reliability

Type of reliability	Description
Repeatability	Where the researchers repeat their study to test and verify their results. If results are reliable, there should be comparable measurements obtained for each successive experiment, provided that they were carried out under the same conditions.
Reproducibility	For any scientific study, an independent researcher should be able to replicate the experiment, under the same conditions, and achieve the same results.
Instrument reliability	A way of ensuring that any instrument used for measuring experimental variables gives the same results every time.

5.1.11 Appropriateness of conclusions and generalisations based on results obtained

In psychological research, when an individual draws a **conclusion** based on the results obtained, they draw an inference as to whether the hypothesis has been supported or rejected. In making this decision, researchers have to ensure that any change in the dependent variable was solely due to the impact of the independent variable, rather than any confounding uncontrolled variables that have not been accounted for and removed in the experimental method.

A conclusion applies only to the sample used in the research. If researchers generalise their results beyond the research, they are stating that their research findings may be applied more widely from the sample tested to the population of interest. A **generalisation** is an application of the conclusions based on the results obtained to a wider population in similar settings outside the study. When generalising, researchers need to ensure that any uncontrolled variables have been accounted for and that the sample size is sufficiently large to be representative of the population. Conclusions derived from analysis of descriptive statistics alone are limited to the investigation data and would only provide tentative generalisations about the wider population.

5.1.12 An overview of ethical considerations in the conduct of psychological research

Ethics are moral principles and standards of practice that are applied by psychologists in order to maintain an appropriate level of care for their clients and/or participants in psychological research. These principles have been formalised into a **code of ethics**, which provides guidelines that psychologists must follow when dealing with humans and/or animals.

The aim of psychology is to provide a greater understanding of behaviour, especially in humans and, in some cases, to use that understanding to predict and control behaviour to improve the human condition.

To achieve this, psychologists often have no other choice but to use human participants to obtain valid results. In so doing, the researcher must take into account additional mental factors (e.g. emotional reactions, embarrassment, guilt, loss of self-esteem) on top of the physical constraints (e.g. pain, stress, anxiety) that are present in research with animals. As humans also have rights of protection and privacy, psychologists must resolve a number of ethical dilemmas.

- Do the 'ends' (knowledge/understanding) justify the 'means' (research design)?
- How far should psychologists be allowed to go in pursuing their knowledge, especially of extreme or undesirable behaviours?
- Will the knowledge gained serve a useful purpose for the betterment of all humankind?

Role of experimenter

In planning research, psychologists must first carefully evaluate the ethical issues involved, including the need to balance the welfare of those who may eventually benefit from the findings of the study against any risk or discomfort for the participants. When working in a multidisciplinary research team, psychologists must make their ethical concerns known to other members and work towards resolving any problems before the study begins.

Beneficence

All psychological research should conform to the overriding ethical principle of **beneficence**, whereby the researcher is responsible for maximising potential benefits of the research and minimising the potential risks of harm or discomfort for all research participants.

Justice

The ethical principle of **justice** should also be applied to promote a fair distribution of benefits and inconvenience within the population of interest, as well as for any individual research participant.

Respect for persons

Researchers should also demonstrate the principle of **respect for persons** by having a proper regard for the welfare, rights, beliefs, perceptions, customs and cultural heritage of all individuals involved in their research.

Non-maleficence

For the principle of **non-maleficence**, the experimenter must take all reasonable steps within the experimental design to *prevent any injury or distress* to the participants.

Informed consent

If the study needs to create physical or mental stress in the participants as a part of the design, the researcher must get the **informed consent** of the participants. In this process, each participant is given all the necessary details about the procedures to be used, including any potential risks and the probable effects to be expected, in order to form an adequate understanding so that they can reach a decision regarding their participation in the experiment. If an individual is incapable of giving valid, informed consent because of mental impairment, permission should be obtained from their guardian or any person or organisation authorised by law to act on their behalf, ensuring that the decision of whether or not to take part is not contrary to the person's best interests. Regardless of this, if the harm or distress is disproportionate to what was expected or deemed acceptable, then the researcher needs to step in to end the experiment and provide appropriate care for the participant.

Deception

Under certain circumstances, it may be necessary for the researcher to use **deception** to hide the research aims from the participants in order to prevent them from behaving differently (e.g. trying to either please the researcher or spoil the results). Many studies would not achieve valid results if deception was not used, and so the researcher must balance the gains against the effect on the participants and ensure that they do not suffer distress as a result of the research procedure.

Debriefing

For studies where the disclosure of the purpose is withheld for scientific reasons, it is essential that **debriefing** of participants occurs to alleviate any distress or remove any misconceptions caused by deception within the experimental design. After the experiment, feedback is given to the participants about the aims and results or interpretations of the study of which they were a part. The aim is to ensure that participants leave the experiment in as similar a state as possible to how they entered it, ensuring that no participant leaves feeling confused, upset or embarrassed. In doing so, steps should be taken to maintain the quality of the relationship with the researcher and to eliminate any mistaken attitudes or beliefs participants may have about the research.

Participants' rights

Confidentiality

Confidentiality is the ethical principle whereby a professional does not disclose to others information given in confidence to them by a client or patient. In research studies, personal information should only be collected if it is relevant to the study and can only be reported in such a way that participants' identities are not revealed. In research using case studies, the anonymity of those involved is preserved through the use of pseudonyms (e.g. 'Sybil Dorsett' or 'Genie') or by using initials (e.g. 'H.M.'). After debriefing, if the participants are unhappy about the protection of their privacy, they can demand that their data be destroyed.

Voluntary participation

Participants also have the right to **voluntary participation** in an experiment; in other words, they have the right to choose to become a part of an experiment if they want to. This choice is often based on being provided with some information as to the purpose, nature and procedures involved in the research design. Participants willingly agree to be involved in the study and are not forced or coerced in any way to participate.

Withdrawal rights

Further to this, **withdrawal rights** enable participants to remove themselves from the research situation at any point (such as when the experiment is seen by the participants as causing too much personal discomfort or distress) and not feel guilty or pressured to continue, regardless of any possible effects on the results.

Despite knowing that they can leave at any point, some participants stay in an experiment even after becoming distressed by the conditions within it. If unexpected severe reactions occur, it is the responsibility of the researcher to immediately terminate the experiment and tend to the needs of the participants, referring them to appropriate services if necessary.

Professional conduct

At all times, the researcher is to act with **integrity**, demonstrating a commitment to act responsibly and follow recognised principles for the honest and ethical conduct and reporting of research, ensuring that the participants' welfare is preserved at all times.

The *APS Code of Ethics* (revised in September 2007) is written by the Australian Psychological Society (APS). Within it, Section A outlines major ethical considerations, while Section B.14 outlines the procedures a psychologist must follow when conducting research. The APS Code of Ethics can be found by searching for it on their site: www.psychology.org.au.

Further to this, psychologists must ensure that procedures followed are in accordance with the National Health and Medical Research Council (NHMRC) guidelines within the *National Statement on Ethical Conduct in Human Research*, 8th edition (2007, updated 2018). This document can be accessed and downloaded by searching for it at: www.nhmrc.gov.au/about-us/publications.

5.1.13 Use of non-human animals in research

Animal studies have contributed to our understanding of many topics in psychology, such as learning theory, neuropsychology, perception, memory formation, parental deprivation and aggression. Such studies have also yielded many practical applications, including behavioural treatments and therapies, along with techniques for training animals for a variety of helpful roles, such as working as guide dogs, home help dogs or drug detection dogs.

Advantages

Psychologists conduct research with animals for several reasons. In many cases, animal studies are performed when it would be *unethical to do such research with humans*; for example, certain forms of brain research or drug testing. This justification is based on the premise that scientists have a primary moral obligation to improve the human condition even, if necessary, through research with other animals.

As animals do not try to understand the purpose of the experiment, their use allows psychologists to conduct experiments that are more *precisely controlled* through manipulation of independent variables and elimination of extraneous variables. They are also seen as more practical, as some animals are easily accessible and their faster breeding cycles allow generational studies to be conducted on the influence of genetics and environmental factors on behaviour.

Psychologists are interested in the study of behaviour, including that of other animal species. Many psychologists, especially the behaviourists, believe that the differences between animals and humans are only quantitative along the *evolutionary continuum*, thereby allowing generalisations to be made from animal research to humans by 'scaling up' the results.

Limitations

Arguments against animal testing in psychology focus on the fact that a number of objective and behavioural measures indicate that animals can be said to suffer stress, pain and anxiety, and that inflicting *suffering* on another creature is morally objectionable. In many cases, the suffering of animals (the means) is often unnecessary and has not been justified by the knowledge gained from the studies (the ends).

Also, humans are *qualitatively different* from other animals. Superficial similarities in behaviour between animals and humans may lead to faulty over-generalisations, while the projection of human-like traits onto animals (anthropomorphism) may also lead to an exaggeration of similarity.

Further to this, laboratory studies on animals are often even more likely to *lack ecological validity* than those conducted on humans, and so these invalid findings are even less useful for generalisation to human behaviour.

Ethical guidelines regarding use of animals in scientific research

To try to resolve such issues, ethical guidelines have been formulated regarding the ethical and humane care and use of animals for scientific purposes, and to prevent the unnecessary use of animal research in psychology. It appears that less psychological research on animals is being conducted today, with animal studies accounting for only a very small percentage of psychological research.

Such codes emphasise that the researchers should ensure that the use of animals is justified, taking into consideration the principle of beneficence by weighing the scientific or educational benefits with the potential effects on the welfare of the animals. Scientific activities using animals should only be performed when they are essential for the acquisition of significant information relevant to the understanding of humans and/or animals and for the maintenance and improvement of their health and welfare.

Researchers who use animals for scientific purposes have a personal responsibility to ensure that the *welfare* of animals is always considered and that their care is directly supervised by a person competent to ensure their comfort, health and *humane treatment*. They have an obligation to treat the animals with respect and to consider their welfare as an essential factor when planning or conducting projects, refining methods and procedures to avoid, or at least minimise, pain or distress in animals used in scientific activities. Any signs of pain or distress not predicted in the design must be alleviated promptly, and such action must take precedence over completing a project. If this is not possible, the animal must be euthanised without delay. If surgery is to occur, the animals must be given the appropriate anaesthesia so they do not experience pain.

Investigators should explore methods to *minimise the number* of animals used in their research, including the development and use of *alternative techniques* that would partially or totally replace the use of animals in scientific activities, such as computer **simulations**. However, the principle of reducing the number of animals used should not be implemented at the expense of greater suffering of individual animals.

Psychologists who have to use animals in their research must ensure that procedures followed are in accordance with the NHMRC guidelines within the Australian Code of Practice for the Care and Use of Animals for Scientific Purposes 2013 (8th edition, updated 2021), which can be accessed and downloaded by searching for it at: www.nhmrc.gov.au/about-us/publications.

5.1.14 American Psychological Association format for reporting psychological research

American Psychological Association (APA) style was developed about 80 years ago and is used to ensure clear and consistent presentation of research material. The use of personal pronouns such as 'you', 'I' and 'we' should be avoided when writing up research. Psychological journals are always written in the third person and in a formal style. The Harvard system of referencing is used where appropriate. Details of APA style are as follows.

Title: Indicates the nature of the investigation

Abstract: A brief summary of the research, which summarises the aim, methods, results and conclusion

Introduction: Background information concerning the nature and reasoning behind the investigation. It may include results of previous research and relevant theory. It includes the aim (the purpose of the topic under investigation) and the hypothesis (a prediction of what the researcher expects to find when empirically testing the effect of the treatment variable). The hypothesis should be fully operationalised (i.e. explained in practical terms).

Method: A step-by-step description of how the research was conducted. This section should be written in the past tense and should read like an instruction manual. This means that a person unfamiliar with the research should be able to read the description and understand how the investigation was carried out.

The method section should be divided under the following subheadings.

- *Participants (subjects):*
    - Who was involved in the study?
    - How were the experimental and control groups selected? (That is, describe the sampling procedure.)

- *Apparatus and materials:*
    - What items were used in the experiment?
- *Procedure:*
    - How was the experiment carried out? What experimental design was used?
    - How was the data collected and analysed?

Results: This section includes clearly labelled summaries of the data that has been selected, statistically analysed and summarised in tabular, graphical, chart or figure form. All tables should be clearly labelled and a sentence explaining them should be provided. One to two sentences analysing and explaining the data should be provided.

Discussion: This section contains a conclusion about whether or not the hypothesis was supported or rejected based on the results obtained. It may also contain a statement on how the results may be generalised. Extraneous variables may be explained, as well as suggestions for improvements to future research designs.

References: Psychology uses the Harvard system of citations and referencing. All books explain this system and provide examples. A common mistake made by many students is to only have a reference list without having any in-text citations to show where these resources have been used. Such citations are usually put at the end of the sentence as '(author(s) surname, year of publication)', unless the author(s)' name forms part of the sentence, in which case the year of publication would be put in brackets immediately after the name. Page numbers are usually only included for direct quotes, although some teachers may request these be included to facilitate the verification of source material. Copying material from other sources without due acknowledgement via citations is treated in academia as plagiarism (a form of cheating), so it is advisable to develop good referencing skills as early as possible.

References should also be clearly presented in the proper format.

- **For books:** Author(s)' surname, Initial(s) (year of publication) *Title of book*. City of publication: Publisher's name.
- **For journal articles:** Author(s)' surname, Initial(s) (year of publication) Title of article. *Title of Journal, Volume number of journal*, page numbers.
- **For Internet sites:** Referencing websites can be difficult, as not all list authors or dates. As with other sources, the aim is to assist in finding the source material if necessary, so include as much information as the online host makes available:
    - **With author:** Author(s)' surname, Initial(s) (year of publication) (online) Title of article. Available from: http://www.fullwebsiteaddress/url (Accessed date)
    - **Where author is unknown:** Title of article. (year of publication) (online) Available from: http://www.fullwebsiteaddress/url (Accessed date)
    - If there is no date of publication or last update on the website, then put the year that you accessed the website.

Appendices: If necessary, any additional background materials (tables, figures, etc.) that do not fit in the above sections, but which clarify the contents of the report, are put after the references and appendices.

5.2 Student-designed practical investigation

In Units 3 and 4, you will design and conduct a practical investigation related to mental processes and psychological functioning. The investigation should relate to ideas and skills developed in Unit 3, Unit 4 or across Units 3 and 4, and may be investigated directly through an appropriate experimental research design involving independent-groups, matched-participants, repeated measures or a **cross-sectional study**.

The investigation requires you to:

- identify an aim
- develop a question
- formulate a research hypothesis including operationalised variables
- plan a course of action that attempts to answer the question and that takes into account safety and ethical considerations
- undertake an investigation either in the laboratory and/or in the field, which involves the collection of primary quantitative data
- analyse and evaluate the data
- identify limitations of data and methods
- link experimental results to scientific ideas
- reach a conclusion in response to the question and suggest further investigations which may be undertaken.

You must also maintain a logbook of your practical work for record, authentication and assessment purposes.

You will then present the methodologies, findings and conclusions of your investigation in a structured scientific poster according to the conventions of scientific report writing and guidelines specified by the Victorian Curriculum and Assessment Authority.

5.2.1 Logbooks

The use of a logbook reflects standard scientific practice. Students undertaking VCE Psychology must maintain a logbook of practical activities in each unit to record all elements of their investigation, including planning, **identification** and management of relevant risks or ethical concerns, recording of raw data, preliminary analysis and evaluation of results. All items in the logbook must be dated and clearly documented.

Not only will the logbook be submitted as a requirement for satisfactory completion of any practical work during the course, but, as part of Unit 4, Area of Study 3, both the scientific poster and logbook entries will be assessed. As a result, all items in the logbook must be dated and clearly documented, and your teachers will regularly check and monitor the logbook during the student-designed research investigations.

Although some schools may allow the logbook to be maintained in electronic form, most schools require that students maintain a hard copy, as is commonly the practice in scientific research, to avoid or reduce the risk of fabrication and/or modification of results for the assessment task.

5.2.2 Scientific posters

Psychologists use posters to present their research at scientific conferences. Work and ideas can be shared and feedback from other researchers can be received in a way that can be less threatening than giving a talk. A poster session at a conference sees researchers display their posters and stand by them to answer questions, which can clarify the research. As with other reporting conventions, posters usually have common headings, which are given in Table 5.4, and should appear in this order.

While formats can vary slightly, posters usually have three or four columns and are read down the columns, from left to right. Columns make each line of text shorter and therefore easier to read.

Posters are not a scientific report placed on a poster; they are different from a written report in that they have much less text and should not be more than 600 words in total. You should see your poster as an illustrated extended abstract.

Graphics and figures are the focus and should tell the story for you; the text is there to support the graphics.

Formatting is important: a carefully selected colour theme (choose two or three colours only) and font theme can be used to create sections in the poster that make it easier to read. All text must be readable from a distance and large blocks of text should be avoided. A font size of at least 20 pt is recommended for text, and section headings should be larger again.

A dark type on a simple, light-coloured background is easier to read. Make sure your poster does not become cluttered and therefore difficult to read. This can be achieved by having 'white space' – sections of space separating your text and graphics.

TABLE 5.4 Presentation format of the research poster

Poster section	Content
Title	A short statement summarising the research question under investigation. This can often be phrased as a research question.
	It needs to attract the reader's attention, be informative and be written at the top of your poster. The title can be bold but should not be all capital letters or all italics.
	The author's name appears immediately below the title in a smaller font. If there is more than one author, names are listed according to their contribution, with the highest contributor listed first. The presenting author's name is underlined.
Introduction	An introduction is a brief explanation or reason for undertaking the investigation, including the aims of the study. It states what you are intending to explore within your investigation followed by a hypothesis predicting your results.
	Any relevant background information, including psychological concepts, models, theories or similar studies, are also outlined and need to be correctly referenced.
	A photograph in your introduction can be used to generate interest and to explain some information so that you use fewer words.
Methodology and methods	This section summarises what you did: the way the experiment was conducted, how ethical and safety guidelines were followed and how data was collected and analysed. This can be done using dot points written in the past tense.
	Diagrams, photographs (e.g. you doing something) or flow charts can be used to describe a procedure with several steps.
	The summary outlining the methodology that you used in your investigation needs to be able to be authenticated from your logbook entries.
Results	Results are presented in a way that will show trends, patterns and/or relationships. Your results, however, should not be interpreted in this section.
	A graph is preferable to a table of raw data, but the type of graph you use must be appropriate. Column and bar graphs are useful for comparing two data sets; although if you want to show a relationship between two variables, a scatter plot might be more appropriate.
	Always make sure you label your axes and choose an appropriate scale so that the data takes up most of the plot area. Avoid using a coloured background on your graph. Grid lines will probably not be necessary; it is the pattern that matters, rather than exact values.
	A legend next to your graph can be used to draw attention to particular features of your results, can refer to the research question and can include relevant statistics, such as a mean.

TABLE 5.4 cont.

Poster section	Content
Discussion	The discussion must interpret and evaluate the analysed primary data and explain what your results mean. Link the results to the research question and state whether or not your results supported your hypothesis. If not, suggest why. Identify the key findings of your experiment and compare them to relevant psychological concepts in existing knowledge, models or theories. Identify unusual results and limitations in the data and methods. Suggest how these limitations could be overcome and what further investigations might be undertaken. Remember that this needs to be done concisely.
Conclusion	A conclusion is a clear statement that responds to the research question on the basis of the data you have gathered. A conclusion should only be a few sentences in length and should not introduce any new information.
References and acknowledgements	All material that has been sourced and used must be referenced. The reference list is a single list at the bottom of the poster, which is alphabetical according to author. Anyone who has helped you in your experiment should be thanked. This includes people who supplied equipment, laboratory assistance, helped develop ideas or assisted you in the interpretation of data.

Once you have drafted your content to express it clearly and succinctly, you can then reorganise it into the required poster format. The poster may be produced electronically or in hard copy format.

The poster should not exceed 600 words. You must be conscious of the word limit through the drafting process and select information carefully so that you stay within the specified limit.

You will have to produce a poster conforming to the VCAA exemplar for a poster presentation, as per the Study Design.

VCAA poster template

The centre of the poster will occupy 20–25% of the poster space and will be a one-sentence summary of the major finding of the investigation that answers the investigation question.

Title Student name		
Introduction	Communication statement reporting the key finding of the investigation as a one-sentence summary	Discussion
Methodology and methods		
Results		Conclusion
References and acknowledgements		

FIGURE 5.7 VCAA template for a scientific poster

VCE Psychology Study Design (2023–2027) © copyright 2022, Victorian Curriculum and Assessment Authority

Checklist for practical investigation

Within the current Study Design, themes relating to research methods will be applied within the practical investigation related to mental processes and psychological functioning undertaken in either Unit 3 or Unit 4, or across both Units 3 and 4. Use the following checklist to ensure you have covered all of the required elements of your investigation.

Tick (✓) the box once you are confident that you have covered each element for your investigation.

Key science skill: Develop aims and questions, formulate hypotheses and make predictions	
Identification of the purpose of the investigation (aim)	
Construction of a research question	
Identification of psychological concepts specific to the investigation and their significance	
Definitions of key terms and psychological representations	
Identification and operationalisation of independent and dependent variables	
Construction of a testable research hypothesis	
Identification of extraneous and potential confounding variables	
Key science skill: Plan and conduct investigations	
Rationale regarding appropriate scientific research methodology and techniques of primary data collection relevant to the selected investigation	
Minimisation of experimental bias and confounding and extraneous variables, including sampling procedures in selection and allocation of participants within an appropriate experimental research design	
Key science skill: Comply with safety and ethical guidelines	
Identification and application of relevant ethical and health and safety guidelines pertinent to the use of human subjects	
Key science skill: Generate, collate and record data	
Generation and accurate recording of primary data	
Organisation and presentation of primary data in a meaningful way, including tables and/or graphs	
Key science skill: Analyse and evaluate data and investigation methods	
Analysis and evaluation of primary data to identify patterns and relationships	
Decisions regarding reliability and validity of data	
Identification of sources of error and limitations of data and methodologies	
Key science skill: Construct evidence-based arguments and draw conclusions	
The key findings of the selected investigation and their relationship to relevant psychological concepts and theories	
Determination as to whether data from the investigation provides weak or strong evidence that supports or refutes a hypothesis or prediction being tested, or whether it leads to a new prediction or hypothesis being formulated	
Discussion of models and theories, and their use in organising and understanding observed phenomena and psychological concepts, including their limitations	
Assessment of the generalisability of statistics from samples to the populations from which the sample was derived	
Key science skill: Communicate and explain scientific ideas	
Psychological report writing and scientific poster presentation according to APA referencing conventions	
Use of appropriate scientific and psychological terminology and representations	
Standard abbreviations and acknowledgment of references	
Use of clear, coherent and cogent expression	
Acknowledgement of all sources of information and assistance, using scientific referencing conventions	

Glossary

accuracy How close a measurement is to the 'true value'. It cannot be a fixed number, but may be described as more or less accurate

beneficence The ethical commitment to maximising benefits while minimising risk and harm when taking a particular position or course of action

case study A type of research investigation that focuses on a particular person or event in-depth. Case studies usually involve direct observation and the gathering of qualitative data and provide insight into a particular psychological phenomenon

code of ethics A set of guidelines outlining standards for professional practice and conduct that must be followed by all psychologists and researchers when working with people and animals

conclusion Part of a research report. A statement that makes a judgement about the meaningfulness of the findings of the investigation. The conclusion should answer the question posed in the aim

confidentiality A participant's right to privacy and security of their personal information, including not being identifiable in the results

confounding variable A type of extraneous variable that ends up changing the dependent variable (DV) in an unwanted way. This confounds the results, as it is impossible to determine the cause of the change in the DV. Confounding variables interfere with the internal validity of the study

control group In a controlled experiment, the group of participants that is not exposed to the independent variable (or treatment). This group provides a comparison for the group that is exposed to the treatment, so ideally, participants are matched to the experimental group on other relevant variables

controlled experiment A type of research investigation where the researcher manipulates one (or more) independent variables and then measures its effect on a dependent variable (DV). This means the researcher must attempt to control the influence of other variables that could also affect the DV. A controlled experiment usually involves the comparison of outcomes for a control group with those of an experimental group

controlled variable (CV) An extraneous variable whose influence has been eliminated from an experiment so that it cannot affect results; it has been controlled using a particular strategy (e.g. matching groups)

A+ DIGITAL FLASHCARDS
Revise this topic's key terms and concepts by scanning the QR code or typing the URL into your browser.

https://get.ga/aplus-vce-psych-u34

convenience sampling Selection of participants because they are readily available to the researcher

correlational study A scientific investigation that involves measuring variables in an uncontrolled (natural) setting to identify and understand any relationships that may exist between them

cross-sectional study Research in which individuals of differing ages (or other relevant characteristic) drawn from a representative sample are compared in a single study

data The observed facts that constitute the results of an experiment

debriefing At the end of a research study, participants are informed of the study's true purpose. This is essential in studies where deception has been necessary. Mistaken beliefs are corrected, and information is provided about services to help with distress resulting from participation

deception Withholding information from participants about the procedures used in and true nature of a study. Used in cases where giving participants the information beforehand might influence their responses and affect the internal validity of the study

dependent variable (DV) The variable that is *measured* in an experiment. It is the variable that is expected to change when exposed to the independent variable (IV). Represented in graphs on the vertical (y) axis

descriptive statistics Statistics used to summarise and organise data. They include measures of central tendency (mean, median, mode), the range and spread (standard deviation) of data, frequency tables and graphs

direct observation A research method involving collection of data by carefully watching and recording behaviour as it occurs

double-blind procedure A procedure where neither the experimenter nor the participants know which of the participants are in the experimental group and are therefore exposed to the independent variable (IV)

empirical evidence Scientific research gathered by the direct method of systematic data collection

ethics Moral principles and values that provide guidance about making judgements and decisions about how to act. Unethical behaviour is that which is morally wrong

experiment A controlled method of data collection used to systematically measure a causal relationship between the independent variable(s) and dependent variable(s) that have been operationalised in a hypothesis

experimental condition The condition in an experiment that contains the independent variable

experimental group In a controlled experiment, the group of participants exposed to the independent variable (IV) – the treatment

experimenter bias An unconscious expectation of the researcher that may influence their observations of data

experimenter effect Any expectations, beliefs or preferences of a researcher that may unintentionally influence their study, the recording of observations or in any way affects the outcome of the investigation

external validity The extent to which the results of an investigation can be applied (generalised) to people or situations beyond the sample

extraneous variable Any variable, other than the independent variable, that may change the results (dependent variable). Researchers try to control extraneous variables before the research starts by thinking of what they could be and then taking steps to stop their effect

fieldwork A data-collecting technique where an animal or person is observed in their natural environment. There is no experimental control of variables. Also known as naturalistic observation

generalisation A decision or judgement about whether results obtained from a sample are representative of the population of interest

hypothesis A testable prediction about the relationship between two variables. It is based on prior knowledge, so it is also considered to be an educated guess

identification The process of recognising phenomena as belonging to a particular category, or as possibly being part of a new or unique set

independent-groups design An experimental research design in which participants are randomly allocated to one of two or more entirely separate (independent) groups, making it equally likely for an individual to be in the experimental or control group

independent variable (IV) The variable systematically *manipulated* by the experimenter to gauge its effect on the dependent variable. Represented in graphs on the horizontal (x) axis

inferential statistics Formal statistical analysis of the data that measures the likelihood of whether results obtained in a study occurred by chance

informed consent The ethical process whereby a research participant is given all the necessary details about the nature and purpose of the experiment, including any potential risks that may be present in the research design, to reach a decision to willingly agree (usually in writing) to participate in an experiment

integrity When completing research, the ethical commitment to searching for knowledge; the honest reporting of all sources of information and results (whether favourable or unfavourable) in ways that permit scrutiny and contribute to public knowledge and understanding

internal validity A measure that ensures the experimental design investigates what it sets out and/or claims to investigate by effectively encompassing all of the steps of the scientific research method, eliminating potential confounding variables in order to determine if a causal relationship exists between the independent variables and the dependent variables

justice The moral obligation to ensure that there is fair consideration of competing claims; that there is no unfair burden on a particular group and that there is fair distribution and access to the benefits of an action

literature review The collation and analysis of secondary data related to other people's scientific findings in order to answer a question or provide background information to help explain observed events or prepare for an investigation to generate primary data

longitudinal study A study that collects data over two or more periods in time, using the same participants. Studies can run for years, or even decades.

matched-participants design (matched-pairs design) An experimental research design that attempts to control subject characteristics by pairing participants according to key characteristics or variables that, if not controlled for may have a confounding effect on the research. Each member of the pair is placed in separate groups (the experimental or the control group)

mean A measure of central tendency that gives the numerical average of all the scores in a data set. It is calculated by adding all the scores in a data set, then dividing the total by the number of scores in the set (see also *median*, *mode*)

median A measure of central tendency, this is the middle score in a data set. It is calculated by arranging scores in a data set from highest to lowest and selecting the middle score (see also *mean*, *mode*)

mode A measure of central tendency, this is the most frequently occurring score in a data set (see also *mean*, *median*)

modelling The construction of either a physical model, such as a small- or large-scale representation of an object, or a conceptual model, which represents a system and how various concepts are related within that system

non-maleficence A research ethic emphasising the avoidance of causing harm. When harm is not avoidable, the harm resulting from any course of action should not be disproportionate to the benefits from that course of action

objective data Data that has been gathered using systematic observation that is not influenced by any personal bias

observation A scientific research method that involves watching and recording behaviour as it occurs in a clinical or naturalistic setting. Data represents an observable event, such as spoken or written responses, test scores and other measures of behaviour

observer bias Bias in results of an observational study that occurs when an observer sees what they expect to see, or records only selected details of an observed behaviour

operational definition The precise, comprehensive description of the concept to be measured in an experiment, and the procedures that will be used to observe, manipulate and/or measure that concept

operational hypothesis The expression of a hypothesis in terms of how the researcher will determine the presence and levels of the variables under investigation; that is, how the experimenter is going to put their hypothesis into operation

outlier Data readings that lie a long way from other results. They may occur by chance or be the result of measurement and recording errors

participants The people or animals used in an experiment to study behaviour, characteristics or responses

participant allocation The systematic process of assigning participants to different groups to ensure that the personal attributes of participants that may influence the results are distributed evenly within the experimental and control groups

p value The probability level that forms the basis for deciding if results are statistically significant (not due to chance)

participant selection The systematic sampling process of choosing participants for research as a representative subset or portion of the population of interest

placebo A medical treatment that looks real, but does not have any active ingredients, so it cannot have an effect on the condition being studied

placebo effect Occurs when a behaviour is caused by an individual's belief or expectation, rather than by a specific procedure that has been administered to elicit that behaviour

population The entire group of people that is of interest to a researcher, from which a sample will been drawn

professional conduct The ethical principle whereby the researcher must act responsibly and ethically to ensure that the participants' welfare is preserved at all times

qualitative measures Factual or descriptive pieces of information about the qualities of the characteristics or behaviours being measured

quantitative measures Numerical measures/values used to quantify or describe the characteristics or behaviours being measured

random allocation A procedure for assigning participants to either the experimental group or control group in an experiment, ensuring that all participants have an equal chance of being allocated to either group

random sampling A sampling technique that uses chance to ensure that every member of a population of interest has an equal chance of being selected to participate in the study

range A measure of spread/dispersion representing the difference between the highest and lowest scores in a frequency distribution

reliability The consistency, dependability and stability of the results obtained from a research study

repeatability Where the researchers repeat their experiment under exactly the same conditions to test and verify their results by checking the level of agreement/consistency between the results obtained in each case

repeated-measures design An experimental research design where the same group of participants undertakes all of the conditions in the experiment; that is, the control (baseline) and the experimental condition(s)

reproducibility The closeness of the agreement between measurements of the same quantity that have been taken under different conditions. These different conditions include a different method, observer, equipment, location or time. When findings are not reproducible, results may lack credibility

respect for persons The ethical principle whereby the researcher should value each participant as a human being and have a proper regard for the welfare, rights, self-determination, perceptions, beliefs, customs and cultural heritage of all individuals involved in their research

sample A group of participants selected to participate in a study, taken from a population of research interest

sampling The process of selecting participants from a population of interest. Sampling techniques can be random, stratified or a sample of convenience

simulation The process of using a model to study the behaviour of a real or theoretical system, especially in cases where the variables cannot be controlled as the system may be too complex, too large or small, too fast or slow, not accessible or too dangerous

single-blind procedure An experiment where participants are unaware of the experimental or control condition to which they have been assigned, but the experimenter is aware of their assignment

standard deviation A measure of the variability of scores in a distribution, indicating the average difference between the scores and their mean.

stratified sampling A sampling technique used to ensure that a sample contains the same proportions from each nominated strata that exist in the population

subjective data Data obtained by self-report measures, where subjects give verbal or written responses to a series of research questions

theory A tentative explanation for a phenomenon that attempts to describe how ideas fit together

uncontrolled variables Variables that have influenced the result, as their presence was not accounted for (and removed) in the experimental method. Uncontrolled variables that cause a change in the value of the dependent variable are termed 'confounding variables'

validity Whether a questionnaire or scale actually measures what it is supposed to measure. The investigation has internal validity when the study produces results that can be interpreted meaningfully in relation to the aims of the study. External validity is achieved when the results can be meaningfully generalised from the sample to the population

variable Any condition (trait, event or characteristic) that can have a range of values. A variable can be manipulated or measured in an investigation

voluntary participation When participants willingly agree to take part in an experiment free from pressure or fear of negative consequences, after understanding what is required of them

withdrawal rights Participants are entitled to leave an experiment (withdraw) for any reason at any stage without any negative consequences. Participants are informed of the right to withdraw before agreeing to participate

Revision summary

Use the following summary of syllabus dot points and key knowledge within Unit 4 Area of Study 3 to ensure that you have reviewed the content thoroughly. Provide a brief definition or comment for each item to demonstrate your understanding or code them using the traffic light system: green (all good), amber (needs some review) or red (priority area to review). Alternatively, write a follow-up strategy.

How is scientific inquiry used to investigate mental processes and psychological functioning?	
Formation of operational hypotheses	
• Hypothesis	
• Independent variable	
• Dependent variable	
• Operational terms	
• Development of an operational hypothesis	
Methodology	
• Case study	
• Classification and identification	
• Controlled experiment	
• Correlational study	
• Fieldwork	
• Literature review	

• Modelling	
• Product, process or system development	
• Simulation	

Participant selection	
• Convenience sampling	
• The strengths and limitations of convenience sampling	
• Random sampling	
• The strengths and limitations of random sampling	
• Stratified sampling	
• The strengths and limitations of stratified sampling	

Participant allocation	
• Control groups	
• Experimental groups	

Extraneous variables	
• The difference between extraneous variables and confounding variables	
• Controlled variables	
• Uncontrolled variables	
• Placebo effects	
• Experimenter effects	
• Single-blind procedure	
• Double-blind procedure	
Research designs used to minimise the effects of extraneous variables in experiments	
• Independent-groups design	
• Matched-participants design	
• Repeated-measures design	
• Critical evaluation of experimental designs	

Data analysis	
• Accuracy	
• Precision	
• True value	
• Random errors	
• Systematic errors	
• Personal errors	
• Uncertainty	
• Outliers	
Descriptive statistics	
• Qualitative data	
• Quantitative data	

Findings	
• Conclusions	
• Generalisations	
• Appropriateness of conclusions and generalisations based on the results obtained by psychological research	
• Internal validity	
• External validity	
• Repeatability	
• Reproducibility	
Ethical principles in the conduct of psychological research	
• The role of the experimenter	
– Non-maleficence	
– Respect for persons	
• The use, protection and security of:	
– Participants' rights	

»	– Confidentiality	
	– Voluntary participation	
	– Withdrawal rights	
	– Informed consent procedures	
	– Deception in research	
	– Debriefing	
	• Professional conduct	
	– Beneficence	
	– Integrity	
	– Justice	
	• The use of non-human animals in research	
APA format for reporting psychological research		
• APA research format		

Exam practice

Investigation design

Multiple-choice questions

Answers start on page 260.

Question 1

A research hypothesis

A states precisely what the outcome of a research study will be.

B is merely an educated guess as to what will happen within the context of the current experiment.

C is a testable statement phrasing the prediction regarding the outcome of a research study.

D anticipates that there will be no significant difference between the variables under study.

Question 2

Within a practical research activity, a psychology class investigated the effectiveness of observational learning. The students were instructed that they were to complete a maze activity as quickly as they could. The students broke into pairs, with one student performing the task while the other watched them while timing their performance. When the first student had finished, they swapped roles and the other student performed the same task. The teacher then recorded the times taken to perform the maze by each group of students.

The study was designed to test the statement that: 'The psychology students who timed first and who were able to observe their partner's performance would take less time to correctly complete the maze in comparison to their partner'. This statement is an example of

A a theory.

B a null hypothesis.

C a research hypothesis.

D an operational hypothesis.

Use the following information to answer Questions 3 and 4.

Parminder compared the effects of consumption of alcohol on reaction times in people at various stages of life.

His stratified sample included participants aged 18 to 70 years. In the repeated-measures experiment, participants consumed one standard drink of alcohol at half-hourly intervals until they reached 0.10% blood alcohol concentration (BAC). Participants completed a series of computer-based tests for reaction times at BACs of 0.00%, 0.05% and 0.10%.

Question 3 ©VCAA 2020 SA Q34

Which of the following includes both an independent variable and a dependent variable for Parminder's study?

	Independent variable	Dependent variable
A	Age	Reaction time
B	Reaction time	BAC
C	Cognitive performance	Amount of alcohol consumed
D	Amount of alcohol consumed	BAC

Question 4 ©VCAA 2020 SA Q33

The graph below represents reaction time, in seconds, versus age, with the lines representing the trend of results for each level of BAC.

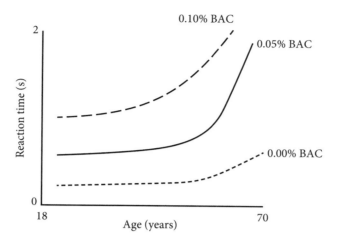

The graph above demonstrates that

A an altered state of consciousness is achieved.
B the higher the BAC, the greater the reaction time.
C an improvement in reaction times occurs as age increases.
D reaction times decrease significantly in both the 0.05% and 0.10% BAC conditions.

Question 5

A major requirement for a sample to be used in psychological research is that it

A comprise a large number of participants.
B be significant.
C be convenient.
D be representative.

Question 6

In order to gain a sample such that gender and age groups are represented in proportion to their numbers within the population, a researcher would need to use a

A random sample.
B stratified sample.
C convenience sample.
D matched-participants design.

Question 7

Research participants are said to be randomly assigned when they

A each have an equal chance of being chosen from the population of interest to the researcher.
B are assigned to experimental and control groups from a sample that is representative of the larger population.
C each have an equal chance of being assigned to either the experimental or control group.
D do not know whether they are in the experimental or control group.

Question 8

In a study into the effects of alcohol on driving ability, the control group should be given

A a dosage of alcohol to bring their BAC up to 0.10.

B half the dosage given to the first group, to bring their BAC up to 0.05.

C a driving test before and after drinking alcohol.

D no alcohol at all.

Question 9

In an experiment, a variable that causes a change in the dependent variable and therefore affects the results of the experiment in an unwanted way is referred to as

A a controlled variable.

B an extraneous variable.

C an independent variable.

D a confounding variable.

Question 10

In an experiment, it is essential to control for extraneous variables so that

A there is a probability that the results will be obtained by chance.

B a valid conclusion can be made about the effect of the independent variable on the dependent variable.

C a valid conclusion can be made about the effect of the dependent variable on the independent variable.

D the hypothesis is supported and the results of the experiment can be generalised to the broader population.

Question 11

As part of an investigation into the effect of caffeine levels on sleep patterns, researchers decide to subject one group of participants to an experimental condition whereby they only consume decaffeinated beverages but are under the impression that they are having a 'normal' amount of caffeine. In so doing, the researchers are

A attempting to eliminate experimenter bias.

B employing a double-blind procedure.

C trying to control for the placebo effect.

D introducing too many variables for analysis.

Question 12 ©VCAA 2018 SA Q36

A researcher was investigating the effects of a gamma-aminobutyric acid (GABA) agonist in the treatment of a specific phobia. Group A, the experimental group, received the GABA agonist. Group B, the control group, received a placebo. Concerned about experimenter bias, the researcher used a double-blind procedure with the help of a research assistant who worked directly with the participants.

Which one of the following identifies the double-blind procedure used in this investigation?

A Only the researcher knew who would receive the placebo.

B Only the research assistant knew who would receive the GABA agonist.

C Only the researcher and the control group knew who would receive the placebo.

D Only the researcher and the research assistant knew who was in the experimental group and the control group.

Question 13

Fifty volunteers agreed to participate in a study to measure whether observational learning was an effective means of developing a new skill. Their results were to be compared with a control group of 50 other volunteers who were only told how to perform the task and who did not receive any demonstration to model the skill. The researchers in this case used

A a double-blind procedure.

B a repeated-measures design.

C a matched-participants design.

D an independent-groups design.

Question 14

Mean, median, mode and standard deviation are all examples of

A descriptive statistics.

B necessary statistical requirements in psychological research.

C nominal scales of measurement.

D ordinal scales of measurement.

Question 15 ©VCAA 2018 SA Q8 (ADAPTED)

A psychologist wanted to investigate people's responses to being pricked by a needle. Details of the investigation were provided to a group of 10 participants prior to the investigation. The investigation involved blindfolding participants and pricking each participant's finger over several trials.

The psychologist repeated the investigation on another group of participants using exactly the same procedure and obtained similar results.

What do the similar results suggest?

A Low validity

B High repeatability

C No confounding variables

D Few participant differences

Question 16

In order to protect the participants' rights to privacy, psychologists apply the ethical principle of

A informed consent.

B debriefing after the research has concluded.

C confidentiality.

D voluntary participation.

Question 17

What must researchers emphasise to potential participants before they participate in a research study?

A The experiment will be explained to the participants, but they are not entitled to feedback about the results of the study.

B Their participation in the study must be voluntary.

C Deception by the experimenter is absolutely not allowed under any circumstances.

D Participants have the right to continue in the experiment if they want to, even if they are exhibiting signs of discomfort or distress.

Short-answer questions

Question 18 (4 marks)

a Distinguish between conclusions and generalisations within research reports. 2 marks

b Identify two factors that should be considered when evaluating the conclusions drawn from a research study. 2 marks

Question 19 (18 marks)

Researchers wish to investigate the effect of caffeine levels on learning and memory. After advertising for volunteers, a random sample was drawn such that it could be split into two equal groups with the same proportions of age and gender as the general population.

A research assistant then supplied unlabelled caffeinated beverages to the first group for their consumption over the 2-week trial and unlabelled decaffeinated beverages to the second group. The latter group, however, was under the impression that they were still having a 'normal' amount of caffeine. Participants monitored their memory for 2 weeks via a diary. After the researchers had finished analysing the data, the assistant informed them which group had consumed the caffeinated beverages.

a Explain why an experiment would be the best choice of research method to investigate this topic. 1 mark

b Name the method employed by the researchers to make their sample more representative of the population. 1 mark

c Identify which experimental procedure was used. Explain the reason behind your answer. 2 marks

d Identify and describe the extraneous variable that this procedure attempts to control within an experiment. 2 marks

e Name and describe the experimental design used in this study. 2 marks

f Outline one of the advantages associated with this type of experimental design. 1 mark

g Outline one of the disadvantages associated with this experimental design. 1 mark

h Identify two possible confounding variables within this experiment. 2 marks

i Identify the experimental group in this study. 1 mark

j State the defining characteristic of an experimental group within a research study. 1 mark

k Outline the purpose of the control group in an experiment. 1 mark

l Name the term describing the condition which the control group was exposed to within this experiment. Briefly explain this type of experimental control, including its purpose in an experiment. 3 marks

Question 20 (25 marks)

Nicole Lamond and Drew Dawson have conducted several studies into the effects of sleep deprivation.

In one study, volunteers were recruited via advertisements placed around local universities. The volunteers kept a general health questionnaire and sleep–wake diary before the study. The following types of participants were excluded: those with current health problems, sleep disorders or a history of psychiatric illness; smokers; those on medication known to interact with alcohol; and those consuming more than six standard drinks per week.

Twenty-two participants aged 19–26 years performed a variety of neurobehavioural tasks under two conditions: (1) in the sleep deprivation (sustained wakefulness) condition, participants were kept awake for 28 hours; and (2) in the alcohol condition, participants consumed an alcoholic beverage every 30 minutes until their blood alcohol concentration (BAC) reached 0.10%.

Performance on the neurobehavioural tasks was measured hourly.

The researchers found that when performance impairment in the intoxicated and sleep-deprived conditions were compared, 17–20 hours of wakefulness produced performance decrements equivalent to those observed at a BAC of 0.05%, and approximately 24–25 hours (one night) of full sleep deprivation produced performance impairment equivalent to those observed at a BAC of 0.10%.

The researchers concluded that moderate levels of sleep deprivation produced performance equivalent to or greater than those observed at levels of alcohol intoxication deemed unacceptable when driving, working and/or operating dangerous machinery.

a   State the main hypothesis being tested. — 3 marks
b   Identify the operationalised independent variables. — 2 marks
c   Identify the operationalised dependent variable. — 1 mark
d   Identify the process used for participant selection. — 1 mark
e   Name the experimental design that was employed. — 1 mark
f   Outline one of the advantages associated with this type of experimental design. — 1 mark
g   Outline one of the limitations associated with this experimental design. — 1 mark
h   Identify and describe the process that would be implemented to try to compensate for the main disadvantage associated with this type of experimental design. Clarify your description by referring back to the study by Lamond and Dawson. — 3 marks
i   Outline the key findings (results) that were derived from this study. — 2 marks
j   On the basis of the results obtained, state whether the hypothesis was supported or not and what conclusions were made by the researchers. — 2 marks
k   What sort of generalisation might the researchers make on the basis of the results? — 1 mark
l   Identify two limitations of the research design and methodology. Explain your answer in each case. — 4 marks
m   Describe any three ethical considerations that needed to be considered within this experiment. — 3 marks

Read the following research study. All the questions which follow relate to this study.

Doctor Fraser is a university research psychologist. His area of expertise is the development of literacy skills in children.

Doctor Fraser has designed a new literacy program for Grade 4 children in Victoria. It is a 30-minute television literacy program that runs daily for 4 weeks.

To test this program, Doctor Fraser sent a letter to all parents/guardians of Grade 4 children in Victoria asking for volunteers. The children of the first 100 parents/guardians who replied were accepted into the study.

Before the experiment, each participant sat a literacy test (Literacy Test A) administered by their Grade 4 teacher. The teachers then sent the results to Doctor Fraser.

Participants were put into 50 pairs based on gender and the similarity of their scores on the literacy test (Literacy Test A).

A computer program was used to select, by chance, one member of each pair to undertake the literacy program. These participants had to watch the literacy program on television for 30 minutes each day for 4 weeks. The other member of the pair was allowed to watch cartoons of their choice for 30 minutes per day for 4 weeks.

At the end of 4 weeks, the participants' Grade 4 teachers administered a second literacy test (Literacy Test B) and sent the results to Doctor Fraser.

Results between the two groups were then compared. A test of significance was calculated and $p > 0.05$.

All ethical guidelines were strictly followed.

Question 21 (2 marks) ©VCAA 2009 E2 SB Q10 (ADAPTED)

Write an appropriate research hypothesis for this study. — 2 marks

Question 22 (2 marks) ©VCAA 2009 E2 SB Q11

Name the independent variable and the dependent variable in this study. — 2 marks

Question 23 (2 marks) ©VCAA 2009 E2 SB Q12

Were the participants in this study randomly allocated? Explain what is meant by random allocation. — 2 marks

Question 24 (4 marks) ©VCAA 2009 E2 SB Q13

a Name the experimental research design that was used in this study. In terms of this study, explain your answer. — 2 marks

b Name one other experimental design. What is one disadvantage of this experimental design compared to the experimental design that Doctor Fraser has used? — 2 marks

Question 25 (2 marks) ©VCAA 2009 E2 SB Q16 (ADAPTED)

Outline one confounding variable and describe how it could affect the results. — 2 marks

Question 26 (2 marks) ©VCAA 2009 E2 SB Q17

In terms of participant selection, should these results be generalised? Explain your answer. — 2 marks

Question 27 (2 marks) ©VCAA 2009 E2 SB Q18

Parents/guardians and participants were debriefed after this study.

Outline two pieces of information that the researcher must give during the debriefing process. — 2 marks

ANSWERS

UNIT 3

Chapter 1 Area of Study 1

Nervous system functioning

Multiple-choice answers

1 D

The nervous system is organised according to two main divisions: the central nervous system (CNS) and the peripheral nervous system (PNS).

A is incorrect because these comprise the CNS. **B** is incorrect because these are structures within the neuron. **C** is incorrect because these are subdivisions of the PNS.

Easy questions like this could occur on the exam. Take each question at face value. Don't assume it must be a trick or read anything else into it.

2 B

The spinal cord connects the brain with the rest of the body.

A is incorrect because the spinal cord is part of the CNS, so it can't send messages to or from itself. **C** is incorrect because movement would be initiated by the motor cortex in the brain. While **D** might imply the spine (or spinal column), it is incorrect because the focus is on the nervous system, not the skeletal system. **D** is actually a function of the cerebellum.

3 C

The two main divisions of the PNS are the autonomic nervous system and the somatic nervous system.

A is incorrect because the afferent nervous system and efferent nervous system comprise the somatic nervous system. **B** is incorrect because, while it extends from the brain to connect it to the rest of the nervous system, the spinal cord is a subdivision of the CNS. **D** is incorrect because these comprise the autonomic nervous system.

4 C

To move your eyes while reading, the motor cortex needs to send impulses via the motor (efferent) pathways, which are a subdivision of the somatic nervous system.

A is incorrect because this division of the nervous system is involved in controlling different levels of arousal. **B** and **D** are incorrect because these are subdivisions of the autonomic nervous system.

5 C

Sensory neurons, not motor neurons, provide feedback from the peripheral nervous system to the central nervous system.

The other options do describe functions of motor neurons, which is contrary to the specific wording of the question.

The key word in the prompt is '**not**'. Usually in multiple-choice questions, you would be looking for the response that is correct. In this case, you are looking for the 'odd one out' that does **not** fit the designated concept. Misreading or only focusing on the latter part of the question would lead to an error in this case.

6 B

Blushing is an autonomic response because it occurs without conscious control and is due to increased blood flow as a result of arousal caused by that particular situation.

All of the other choices are not appropriate for the prompt as they involve different aspects within the CNS. (**A** pertains to the spinal reflex, **C** is a function of the cerebellum and **D** involves areas of the cerebral cortex.)

7 C

The autonomic nervous system controls the activity of the glands and internal organs to enable a balanced state of homeostasis.

A is incorrect because the autonomic nervous system (ANS) is not part of the central nervous system (CNS). **B** is incorrect because relaxation techniques, meditation and biofeedback can be used to enable conscious control of autonomic functions. **D** is incorrect as the ANS affects the visceral muscles and does not 'relay' messages between the CNS and the voluntary muscles. Also, voluntary muscles do not control our internal organs and glands.

8 C

The fight-or-flight response is a state of physiological arousal controlled by the sympathetic nervous system.

A is incorrect because this response is due to the activation of a branch of the autonomic nervous system. **B** is incorrect because this response is automatic and not due to conscious activation by the CNS. Even though this response occurs due to a stressful situation, **D** is incorrect because the changes described are not caused by the parasympathetic nervous system.

While the current accepted terminology for this state would be the fight–flight–freeze response, the wording is appropriate in this case to clearly distinguish between the relative dominance of the autonomic systems in the fight–flight–freeze response – **C** being dominant for fight-or-flight (as required by the prompt) and **D** being more dominant for the freeze component of the response.

9 D

Being startled would activate Jacob's sympathetic nervous system, which would lead to an increase in adrenaline and reduced movement in the large intestine.

A is incorrect because airway passages dilate when the sympathetic nervous system is activated. **B** is incorrect because salivation decreases due to inhibition of digestion when the sympathetic nervous system is activated. **C** is incorrect because the release of cortisol does not occur in the fight–flight–freeze response.

10 D

Once a situation that has triggered the fight–flight–freeze response has been resolved, the parasympathetic nervous system is activated to bring our level of arousal back down to normal levels.

A is incorrect because it was activated to increase the heart rate and blood pressure in response to the perceived threat. **B** is incorrect because it is involved in voluntary actions, such as walking, not changes in internal organs. **C** is incorrect because it is too general and includes the other subdivision that was not active at that stage.

11 B

A conscious response involves awareness of the stimulus and voluntary control by the nervous system.

A is incorrect because conscious responses are voluntary. **C** is incorrect because unconscious responses are involuntary. **D** is incorrect because an unconscious response is not always regulated by the autonomic nervous system (e.g. the spinal reflex).

Absolutes (such as always, never etc.) are seldom used in Psychology, so any statements using such phrasing should only be considered if totally true, which would be rare.

12 A

This question requires acknowledging the scenario as pertaining to the spinal reflex (or reflex arc). Such a response does not involve the brain directly, even though it does become aware of it after the fact.

The key word in the prompt is 'except', requiring you to look for the option that does not apply to the specific behaviour described.

13 B

B is correct because neurotransmitters convey a message to a neighbouring neuron by travelling across the synaptic gap from the axon terminals to the next neuron. The wording of **B** pertains to the neural impulse within the neuron.

All of the other statements are true of neurotransmitters, and are therefore incorrect.

The key words in the prompt are 'not true', requiring you to identify the option that is **incorrect** in relation to the designated concept.

14 C

Synaptic vesicles in the axon terminal release neurotransmitters into the synaptic cleft, and they lock into receptor sites of receiving dendrites.

This question requires an understanding of structural and functional aspects of neuronal communication. **A** is incorrect because communication between neurons is chemical (via neurotransmitters), not via an electrical charge (which pertains to the action potential within the neuron). **B** is incorrect because neurons do not physically connect but are separated by the synaptic gap. **D** is incorrect because the mechanisms involved are known and well-documented.

15 D

Neurotransmitters affect the postsynaptic neuron by causing a change in the electrical potential, either exciting or inhibiting the next neuron.

A is incorrect because they do not stay in the synapse, but either connect with receptors on the postsynaptic neuron or are taken back into the synaptic vesicles (reuptake). While correct in describing the different types of neurotransmitters, **B** and **C** are not be appropriate for the prompt as they are too specific and do not account for the other type of response in the postsynaptic neuron. **D** is therefore the best option as it is the most complete answer.

16 D

As the main excitatory neurotransmitter throughout the brain and central nervous system, glutamate plays a key role in synaptic plasticity by initiating activity in postsynaptic neurons.

A is incorrect because this is a function of the glands. **B** is incorrect because neurotransmitters do not go along the axon. **C** is incorrect because neurohormones travel through the bloodstream, not across the synaptic gap.

17 C

Inhibitory neurotransmitters act to make it more difficult for the next neuron to fire.

A is incorrect because they make it harder for a postsynaptic neuron to generate an action potential. **B** is incorrect because they act to counter the effects of excitatory neurotransmitters, which decreases the likelihood that the postsynaptic neuron would be activated. **D** is incorrect because there is a correct response available.

The temptation for questions like this is to go for **D**, especially if you are rushing. Just because an option combines several choices together does not make it more correct. For you to accept this option, you should read **all** of the options carefully to check that they all are in fact incorrect.

18 A

Neuromodulators produce a slow and long-lasting effect through prolonged activation of target cells.

B is incorrect because neuromodulators are not reabsorbed by the presynaptic neuron but remain in the synapse. **C** is incorrect because neuromodulators can, via second messengers, indirectly affect postsynaptic targets quite far away from the point of release. **D** is incorrect because neuromodulators act through volume transmission, affecting a diverse group of postsynaptic targets.

Just because multiple-choice options are longer and may contain jargon does not automatically mean that they are the correct answer. The simplest answer in this case was the appropriate choice.

19 B

Dopamine is associated with voluntary movement, attention, motivation and pleasure/reward seeking.

A is incorrect because the functions listed relate to noradrenaline. **C** is incorrect because this is performed by acetylcholine. **D** is incorrect because this pertains to GABA.

20 A

Serotonin plays a major role in mood, anxiety and happiness, and is also involved in sleep, memory processing and appetite.

B is incorrect because dopamine is associated with voluntary movement, attention, motivation and pleasure/reward seeking. **C** is incorrect because noradrenaline is associated with arousal, concentration, mood elevation and short-term memory. **D** is incorrect because glutamate stimulates neural activity and is associated with long-term potentiation.

21 C

Long-term potentiation describes the strengthening of neural connections as a result of repeated stimulation.
A is incorrect because long-term potentiation strengthens communication across synapses. **B** is incorrect because long-term potentiation involves an increased release of glutamate. **D** is incorrect because increased levels of this inhibitory neurotransmitter are counterproductive to the process of long-term potentiation.

Questions on the exam will often link dot points across areas of study from Unit 3 and/or Unit 4. This question links an understanding of long-term potentiation with classical conditioning and the physiological basis for memory, both of which are covered in the next area of study.

22 B

A thicker and heavier cerebral cortex was due to the process of sprouting, whereby additional branches grew out from each neuron to create more new synapses with other neurons.
A is incorrect because the higher levels of acetylcholine are associated with increased transmission due to neural stimulation and the formation of memory, but they do not cause the neurons to become bigger. **C** is incorrect because the brain cannot produce new neurons once it has been formed. **D** is incorrect because density would increase because of more branches, not fewer.

23 A

Increased release of glutamate from the axon terminal strengthens synaptic connections in long-term potentiation.
B is incorrect because long-term potentiation involves an increase in the sensitivity of postsynaptic receptors, not a decrease. **C** is incorrect because long-term potentiation creates new synapses. As other choices are invalid, **D** is incorrect.

24 D

In relation to neural plasticity, long-term depression refers to selective weakening of specific synapses in order to make constructive use of synaptic strengthening caused by long-term potentiation (LTP).
A is incorrect because this description applies to mental illness, not neural plasticity. **B** is incorrect because long-term depression occurs due to low frequency or lack of stimulation. **C** is incorrect because long-term depression involves a decrease in the number of glutamate receptors on the postsynaptic neuron, resulting from a lack of stimulation.

Short-answer sample responses

25 a The central nervous system (CNS) regulates, coordinates and controls the major functions of the body along with reflexive processing. (1 mark) The CNS consists of the brain and the spinal cord. (1 mark)

   b • To pass sensory information from the peripheral nervous system to the brain (1 mark)
   • To transmit information from the brain to the peripheral nervous system (1 mark)
   • To control and allow the reflex arc (1 mark)

26 a The peripheral nervous system (PNS) branches out from the CNS and carries information to or from the rest of the body. (1 mark)

> Because this question is worth only 1 mark and the prompt specifies brevity, the challenge is to be clear and succinct about essential aspects, rather than write an essay to cover every detail, especially as there are such prompts in part b of the question.

   b The somatic nervous system:
   • carries messages from sensory receptors to the central nervous system (1 mark)
   • transmits messages to control voluntary movement of the skeletal muscles (1 mark).

The autonomic nervous system:
- connects the central nervous system with the internal organs and glands (1 mark)
- controls the arousal and subsequent relaxation of these organs (1 mark).

> 2 marks for describing the role each branch plays in controlling bodily functions (× 2); each of the main branches of the PNS must be clearly identified in order to specify the differences required by the prompt

27 a The sympathetic nervous system is responsible for creating a state of arousal that mobilises the body's energy and resources during times of stress and arousal, thereby mediating bodily functions involved in the fight–flight–freeze response to stress. (1 mark) A sympathetic response dilates pupils, inhibits salivation, relaxes airways, accelerates the heartbeat, inhibits digestion, and so on. (1 mark)

b The parasympathetic nervous system is concerned with conserving the body's energy and resources during relaxed states. (1 mark) A parasympathetic response constricts pupils, stimulates salivation, constricts airways, slows the heartbeat, stimulates digestion, and so on; hence it balances out the sympathetic response to return the body to a state of homeostatic equilibrium. (1 mark)

> The prompt asked for examples (plural) but did not specify how many. If this is the case in the exam, three appropriate examples should be sufficient.

28 a The fight–flight–freeze response is an automatic response that prepares an animal to deal with a threatening situation by confronting it (fight), which the gazelle is not able to do, by running away (flight), which the gazelle attempted initially and later in order to escape, or by reducing the exposure to the perceived threat (freeze), such as the gazelle playing dead until the chance to escape presented itself. (2 marks)

> 1 mark for describing the different aspects of the fight–flight–freeze response and 1 mark for appropriate references to the scenario to clarify each aspect.
> Be careful not to use circular expression (i.e. using words as part of their own definition), otherwise you would not receive any marks for your answer.
> In this case, you would need to paraphrase the terms 'fight', 'flight' (or 'flee') and 'freeze' in order to demonstrate your understanding of these concepts.

b When an animal is under threat, the sympathetic nervous system is immediately activated, triggering a release of adrenaline that causes an increase in certain bodily functions to enable a quick response, to fight or run away, and inhibit non-essential functions. (1 mark) If the threat is perceived to be too intense, the 'freeze' response may be initiated where the parasympathetic nervous system takes over to overshadow the existing effects of the sympathetic nervous system activation in order to prevent movement to reduce the exposure to the perceived threat, such as playing dead, and conserve energy until there is a chance for escape. (1 mark) When the threat has ended, the parasympathetic nervous system counteracts the effects of arousal by slowly calming the body, restoring physiological systems to a balanced, homeostatic state. (1 mark)

c The fight–flight–freeze response is an adaptive survival mechanism that enables us to react effectively to a threat and helps minimise harm to our wellbeing to enhance our chances of survival, and thus live on and reproduce. (1 mark) The fight–flight–freeze response produces physical responses of autonomic arousal to allow a person or animal to deal with a threat by confronting it (fight), by running away (flight) or by reducing the exposure to the perceived threat by not drawing attention to itself (freeze). (1 mark)

29 a As it was an intentional action, Meera reaching for her coffee mug was a voluntary/conscious response (1 mark) involving the somatic nervous system (1 mark).

b i The spinal reflex (1 mark)

> The following terms have also been accepted in past exams: reflex/reflex arc/reflexive response.

ii Sensory neurons transmit the neural impulses from the sensory receptors to the spinal cord. (1 mark) Interneurons relay a neural impulse to the motor neuron. (1 mark) The motor neuron sends a signal to the skeletal muscles in order to pull back the hand. (1 mark) The brain is not directly involved in the response but registers the sensory input independently of (after) the reflexive pulling back of the hand. (1 mark)

30 a Neurotransmitters are chemicals that alter the electrical activity in other neurons (1 mark) after they are released into the synaptic cleft and attach to receptor sites on the soma or dendrites of the next neuron (1 mark).

b Different neurotransmitters have differently shaped molecules. (1 mark) Each neurotransmitter molecule acts like a key. (1 mark) This key will only fit into its specific complementary receptor site on the postsynaptic neuron, which can be thought of as a lock. (1 mark)

31 a Excitatory neurotransmitters increase the likelihood of the postsynaptic neuron firing in action potential (1 mark), whereas inhibitory neurotransmitters decrease the likelihood of postsynaptic potentials (1 mark).

b As the main excitatory neurotransmitter throughout the brain and central nervous system, glutamate enhances information transmission by initiating activity in postsynaptic neurons. (1 mark) During learning, repeated stimulation of the postsynaptic neuron will increase the number of glutamate receptors on the cell and can initiate growth of new dendritic spines on postsynaptic neurons, resulting in the formation of additional synaptic connections. (1 mark)

32 a Neuromodulators are chemical messengers that act together with neurotransmitters to enhance the excitatory or inhibitory responses of receptors (1 mark) to generate a longer, more-sustained signal that can lead to lasting changes in cellular activity (1 mark).

b Any three of the following (3 marks):

- Neurotransmitters act within a specific synapse, thereby affecting one or two postsynaptic target neurons or effector cells, whereas neuromodulators go beyond the specific synapse and can affect a diverse group of postsynaptic neurons or effector cells.
- Neurotransmitters are received by adjacent postsynaptic targets, whereas the site of action for neuromodulators can be either near or quite far away from the point of release.
- Neurotransmitters act directly on their postsynaptic target by binding to specialised receptors to cause a specific response, whereas neuromodulators indirectly affect their postsynaptic targets via second messengers.
- Neurotransmitters are degraded or rapidly recycled by the presynaptic neuron (reuptake), whereas neuromodulators are not degraded rapidly or reabsorbed by the presynaptic neurons.
- Neurotransmitters produce a rapid, short-lived effect to pass the signal from one neuron to another, whereas neuromodulators enable prolonged activation of target cells that produces a slow and long-lasting effect to change the cellular or synaptic properties of neurons.

33 a Learning is a function of establishing new synaptic connections that are brought about by stimulation through activity, leading to changes in the structure of existing neural networks. (1 mark)

b At the synapse, nerve impulses are converted to chemical processes that either:

- excite activity in the connecting neuron, thereby strengthening and possibly creating new synaptic connections (1 mark), or
- inhibit activity in the connecting neuron, weakening the connections and leading to the deterioration of existing synapses (1 mark).

c Long-term potentiation (LTP) refers to the long-lasting strengthening of synaptic connections of neurons. (1 mark) This results in the enhanced or more effective functioning of the neurons whenever they are activated, thereby improving the ability of two neurons to communicate with one another at the synapse. (1 mark)

d The structure of neurons is changed by sprouting more interconnecting branches (dendrites and axon terminals). (1 mark) This results in more connections or communication points (synapses), allowing more pathways to quickly transmit messages between neurons. (1 mark)

e Long-term depression (LTD) involves the selective weakening of specific synapses due to low frequency or a lack of stimulation. (1 mark) This results in a decrease in the number of glutamate receptors on the postsynaptic neuron, which could prune the number of dendritic spines, resulting in reduced synaptic connections. (1 mark)

34 a Sprouting is the process whereby neurons grow more branches (dendrites and axon terminals) following stimulation and repeated activation in order to make more connections (synapses) with other neurons. (1 mark)

b If damage has occurred within the brain and communication pathways are lost, surviving neurons may seek out and connect with other neurons, rerouting messages to be sent along a new neural pathway to go around the damaged area. (1 mark)

c If a synapse is not regularly activated, the connection will weaken and ultimately be removed through the process of synaptic pruning. (1 mark)

Stress as an example of a psychobiological process

Multiple-choice answers

1 B

A stress reaction occurs whenever a situation is perceived as a threat.
A is incorrect because stress involves the autonomic nervous system. **C** is incorrect because stress is associated with the action of the sympathetic nervous system. **D** is incorrect because the somatic nervous system is not involved in the stress reaction.

2 A

Physiological changes such as dryness of the throat and mouth, pounding heart and trembling are associated with the release of adrenaline into the bloodstream as part of the flight–fight–freeze response.
B and **C** are incorrect because stress is associated with the activation of the sympathetic nervous system, not the parasympathetic nervous system. **D** is incorrect because the effects described are due to adrenaline, not pituitary hormones.

3 D

The gut–brain axis (GBA) involves bidirectional (two-way) communication between the central and enteric nervous systems.
A is incorrect because the brain regulates gastrointestinal function. **B** and **C** are aspects of the gut–brain axis, but are incomplete, especially given **D**.

4 C

As a part of the gut–brain axis, the gastrointestinal microbiome interacts directly with the CNS by influencing brain chemistry and affecting neuroendocrine systems associated with the stress response, anxiety and memory function.
A is incorrect because this applies to serotonin, not dopamine. **B** is incorrect because it is too narrow and does not address the effect the microbiome has on the brain. **D** is incorrect because the composition of this microbial community changes over time and is susceptible to both external and internal factors.

5 B

Cortisol increases glucose in the bloodstream and reduces inflammation.

A is incorrect because cortisol does not stop the immune system functioning, it suppresses it. **C** is incorrect because this is not a function of cortisol. **D** is incorrect because cortisol is not associated with the 'initial alert' but is released after prolonged stress.

6 B

Continued cortisol release during the resistance stage would weaken Jamie's immune system, resulting in his body being unable to fight the cold.

A is incorrect because 'shock' is part of the 'alarm reaction' stage, not a stage in itself within the General Adaptation Syndrome model. **C** is incorrect because exhaustion is more applicable to the resistance stage, as in the exhaustion stage the individual would be prone to more serious illnesses than catching a cold. **D** is incorrect because adrenaline is released during the countershock phase of the 'alarm reaction' stage, and it is cortisol that suppresses the immune system.

7 D

The General Adaptation Syndrome model explores our physiological response to ongoing stress, in particular the involvement of the autonomic nervous system (making **A** and **C** incorrect).

While the parasympathetic nervous system is implicated for responses in the shock stage, the sympathetic nervous system is particularly active in raising our physiological levels to deal with the stressor during countershock. As the alarm reaction phase incorporates both stages, **B** would be incomplete, making **D** the correct answer.

8 C

Hoana is **most** probably in the resistance stage of the General Adaptation Syndrome.

A is incorrect because the demands have increased and so the stress levels, and symptoms, would still be present, requiring Hoana to continue making adjustments to her lifestyle. Just because someone may be under high levels of stress does not automatically mean that they would develop a mental disorder (**B**) and as she is still able to function, it would appear that Hoana has not yet exhausted the resources she needs to deal with her stress. Denial (**D**) could be present in the resistance stage, but Hoana may be fully aware of how she is reacting to the stressors, just not willing to complain to others about it.

9 A

The stage of exhaustion is the stage of the General Adaptation Syndrome in which bodily resources are depleted, and may result in a serious health loss or major collapse.

The other options are incorrect because the person would still have sufficient resources to deal with the stress.

10 B

When undergoing prolonged stress, the immune system may be suppressed resulting in harmful cells not being detected and therefore not eliminated quickly enough, which may result in illness.

Therefore, **C** and **D** are incorrect because both options refer to an increase in immune system activity when undergoing prolonged stress. **A** is incorrect because results indicate a strong correlation between stress and disease, which is not the same as definitively proving that stress causes these conditions. Stress may increase the likelihood of contracting a condition such as heart disease, but it is not an automatic consequence.

11 B

Stomach ulcers are a physiological effect of prolonged arousal, whereas aggression is a psychological effect.

Common errors for this type of question stem from either misreading the question and/or confusion between physiological and psychological bases of behaviours.

A and **C** are incorrect because the answers are in the opposite order to the prompt. **D** is incorrect because both are physiological effects of prolonged arousal.

12 D

This question requires an understanding of different concepts in Lazarus and Folkman's Transactional Model of Stress and Coping.

Questions like this are common in texts and on websites, but you should **not** automatically assume that 'All of the above' will apply in every case – in fact, this is often not the case in VCAA examinations. For you to accept this option, you should check that **all** the options are in fact correct. If one of the options is wrong, then **D** will automatically be wrong, leaving you with two options from which to choose.

13 A

As Terri was upset by her predicament, she had appraised the situation as a threat.

B and **D** are incorrect because these would lead to a positive, confident response that does not fit Terri's reaction. **C** is incorrect because it is obvious that it is very relevant to Terri.

14 C

In Lazarus and Folkman's Transactional Model of Stress and Coping, primary appraisal evaluates the situation, whereas the secondary appraisal evaluates resources for coping.

A and **B** are incorrect because both appraisals involve conscious evaluations. **D** is incorrect because decisions about coping strategies occur as a result of secondary appraisal.

15 D

One advantage of Lazarus and Folkman's Transactional Model of Stress and Coping is that it focuses on a person's psychological response to the stressor, rather than just looking at the physiological effects.

A is incorrect because it is difficult to explore the processes through experimentation. **B** is incorrect because it does not focus on physiological responses. **C** is incorrect because cognitive evaluations form the crux of the model.

16 A

Denial is a maladaptive method of coping because it causes the individual to avoid the problem rather than handle it.

The other options are adaptive strategies that can help someone manage and deal with a problem.

17 D

Rose used both an avoidance strategy (leaving the studio) and an approach strategy (returning to the studio and studying her notes).

A and **B** are incorrect because they each only address one aspect of the scenario. **C** is incorrect because exercise was not involved.

18 B

A and **D** are incorrect because coping flexibility was evident when Theodore changed his strategy by going to the careers counsellor. **C** is incorrect because Theodore is using an approach strategy, not avoidance.

Short-answer sample responses

19 a Stress involves prolonged arousal due to the body's response to threatening or challenging events. (1 mark) For example, a rise in heart rate and blood pressure, a rise in adrenaline levels, higher conductivity of electrical impulses across the skin, and so on. (1 mark) Stressors, on the other hand, are circumstances that are perceived as putting demands on one's physical and/or psychological wellbeing; in other words, the causes of stress. (1 mark) For example, impending exams, relationship problems, major life changes, and so on. (1 mark)

   b **i** The sympathetic nervous system (1 mark)

      ii The hypothalamic–pituitary–adrenal (HPA) axis (1 mark)

20 Any two of the following:
- Release of adrenaline would make Annie more alert and able to focus on the road.
- Redirecting blood supply to the muscles would enable quicker reflexes and better motor control.
- Dilated pupils would enable Annie to take in more visual information and anticipate hazards.
- Increased glucose secretion would give Annie more energy.

> 1 mark for each correctly identified physiological aspect of the stress response (up to a maximum of 2 marks); 1 mark for providing a correct reason as to why this response could be helpful in performing the driving test (up to a maximum of 2 marks)

21 a The gut–brain axis (GBA) is the two-way connection and communication pathway between the gut microbiome and the brain. (1 mark) Your brain monitors and integrates autonomic gut functions to ensure the proper maintenance of gastrointestinal homeostasis and mechanisms such as immune activation. (1 mark) Conditions in the gut microbiome can directly influence brain function and can have differing effects on mood, motivation and higher cognitive functions. (1 mark)

b Through the GBA, gut microbiota can influence brain chemistry and affect neuroendocrine systems associated with the stress response, anxiety and memory function. (1 mark) Gut microbiota produce many of the neurotransmitters required to regulate mood, aid cognition and regulate our body clock. (1 mark)

c Activation of the sympathetic nervous system during chronic stress redirects blood flow and oxygen from the stomach and gastrointestinal system, contributing to indigestion. (1 mark) Chronic stress also activates the hypothalamic–pituitary–adrenal (HPA) axis to release cortisol from the adrenal glands. (1 mark) Increased cortisol levels cause changes in the gut microbiota and intestinal epithelium that can affect the body's immune system and its ability to fight harmful bacteria in the gut, which might be contributing to Will's indigestion. (1 mark)

22 a In the *alarm reaction* stage, a person or animal recognises a threatening situation. In the *shock phase*, resistance to stress drops below normal levels, lowering blood pressure and body temperature, and affecting muscle tone. The *countershock phase* occurs when the sympathetic nervous system activates, giving rise to hormonal and physiological changes within the body in order to respond to the stressor.

> As the prompt specified the use of appropriate terminology, the italicised terms must be used in each case. 1 mark for a general description of the alarm reaction stage; 1 mark each for describing each phase within this stage

b During the resistance stage, physiological arousal stabilises at a point that is higher than normal, and it appears as if the individual is coping with the stress. (1 mark) However, the ability to cope with additional stressors is lowered, and psychosomatic symptoms begin to occur due to the sustained release of cortisol. (1 mark)

c During the exhaustion stage, the body's resources get used up and the ability to maintain physiological arousal significantly decreases. (1 mark) During this stage, organisms are more susceptible to disease, which can lead to serious health problems. (1 mark)

d
- The *shock* phase of the *alarm reaction* stage occurs when a student gets their list of work to be done and/or their list of assessment dates at the start of the year. This would begin their stress reaction as the list(s) may seem overwhelming at that point in time, and as there is often more pressure on them to do well in their final year of school.
- In *countershock*, resources are brought to bear in order to get organised to meet the demands of their Year 12 workload.
- The *resistance* stage occurs when the student tries to balance their workload in order to get their assignments done while revising for their exams (in addition to any other demands that may have been placed on them at home, in their social life, at part-time jobs or by extracurricular activities such as sporting teams or other involvements).

- Within this stage, the stress is still present, but the focus on getting the job done is the student's way of dealing with it.
- The *'exhaustion'* stage occurs when the student gets tired because their body has been in a state of increased arousal/stress for a long period of time.
- Within this stage, the student is more susceptible to illness because the immune system is suppressed by the prolonged release of cortisol during long periods of stress.

1 mark for each dot point that correctly describes the essential characteristics of each stage

23 ©VCAA 2017 MARKING GUIDELINES SB Q1 (ADAPTED)

a Resistance (1 mark)

b Alarm reaction (or alarm, or alarm countershock) (1 mark)

As the prompt specified the stage (and not the phase), 'countershock' is not accepted if 'alarm' is not also mentioned.

c Exhaustion (1 mark)

24 a Avantiika would be in the exhaustion stage (1 mark) because she is constantly getting sick and is unable to meet her deadlines (1 mark).

b The first signs of illness are likely to appear during the stage of resistance. (1 mark)

c Chronic (ongoing) stress increases the levels of cortisol in the bloodstream. (1 mark) This in turn weakens or suppresses the effectiveness of the body's immune system. (1 mark) Lowered or weakened immunity means that the individual is unable to fully fight/defend themselves against harmful cells (or infectious diseases/viruses), thereby increasing the likelihood of illness (or sickness). (1 mark)

25 a According to Lazarus' Transactional Model of Stress and Coping, stress is a function of the discrepancy between perceived demands of the situation and the person's resources for meeting those demands. (1 mark) This means that the person's psychological appraisal of the situation and resources are critical for determining whether the person experiences stress and shows a stress response. (1 mark)

b According to the Transactional Model of Stress and Coping, stress depends on the subjective appraisal or evaluation of environmental events. (1 mark) If people believe that a challenge (good or bad) will tax or exceed their resources, they experience stress. (1 mark)

c Primary appraisal involves assessing whether the stressor is irrelevant, benign-positive or stressful. (1 mark) If it is considered stressful, the person judges it (1 mark) in terms of:
- harm-loss (how much damage has already occurred) (1 mark)
- threat (expectation of future harm) (1 mark)
- challenge (an opportunity for personal growth) (1 mark).

d Eustress is 'good' stress that is considered healthy and necessary to keep us motivated, alert and focused (1 mark), such as benign-positive or challenge state (1 mark). Distress is perceived as causing negative effects on the body due to heightened levels of arousal and is therefore considered unhealthy (1 mark), such as harm-loss or threat (1 mark).

e Having assessed a situation as being stressful, secondary appraisal is an evaluation of the resources available for meeting the potential threat. (1 mark) If the demands required to deal with the situation greatly exceed the resources available, a large discrepancy will be perceived, and the person will experience stress as a result. (1 mark)

26 Strengths
- Focuses on psychological causes of the stress response
- Emphasises the subjective nature and individuality of the stress response
- Regards stress as an interaction between a person and their external environment in which the individual has an active involvement

- Acknowledges personal evaluations of a situation, thereby considering how the individual perceives the situation
- Explains why different people respond in different ways to the same stressors
- Allows for the fact that stressors and people's responses to them can change over time
- Allows individuals to re-evaluate their thinking about a stressor and change their response
- Proposes different strategies for coping with psychological responses to stressors

Limitations
- The subjective nature of individual responses to stress are difficult to test through experimental research
- Individuals may not always be aware of all the factors behind their stress response
- Individuals can experience a stress response without consciously appraising a situation or event
- Neglects physiological responses
- Primary and secondary appraisals can combine with one another and often occur at the same time
- Different types of appraisals can be difficult to isolate as separate variables for study

> 1 mark for each strength (up to a maximum of 2 marks); 1 mark for each limitation (up to a maximum of 2 marks)

27 a Coping strategies refer to the behavioural and psychological efforts that people employ to master, tolerate or minimise their response to a stressor. (1 mark)

b A coping strategy may be applicable and successful in certain circumstances but could be ineffectual or possibly detrimental in other situations. (1 mark) A coping strategy has context-specific effectiveness when it is an appropriate and adaptive choice to deal with a given stressor. (1 mark)

c Coping flexibility involves the ability to choose a coping strategy that is appropriate for the stressful situation (1 mark) and successfully change or adapt it to meet the changing requirements of the situational dynamics (1 mark).

28 a Approach strategies for dealing with stress involve planning and applying techniques to actively tackle the situation directly, consequently reducing the stress by changing or eliminating its source. (1 mark) Avoidance strategies for dealing with stress focus on detachment or withdrawal from the stressor, with no effort to tackle the situation or its causes. (1 mark)

b Benefits
- Generally considered to be more adaptive and effective
- Necessary when action is needed to deal with a serious problem
- Use of approach strategies is usually associated with better outcomes, fewer psychological symptoms and more-effective functioning
- Enable the person to look for ways of managing or removing the problem to provide a long-term solution

Limitations
- Difficulty coping with a number of stressors at the one time
- Ineffectual when the stressor is something that cannot be changed or modified
- Ineffective when the person does not feel that they have any control over their problem

> 1 mark for each benefit (up to a maximum of 2 marks); 1 mark for each limitation (up to a maximum of 2 marks)

c   Benefits
   - Avoidance strategies may enable short-term coping, especially if the stressor is too intense to deal with straight away.
   - Detachment might be appropriate when the source of stress is outside the person's control (such as waiting for the results of an exam).
   - If there are several issues to be dealt with concurrently, selectively avoiding certain stressors, especially if they are too complex, would enable the individual to focus on matters that could be resolved to effectively respond to the larger problems at a later stage.

   Limitations
   - Avoidance strategies tend to be maladaptive.
   - Prolonged use can hinder people from dealing with stressors in productive ways.
   - A delay in dealing with a stressor can be harmful when a response is needed right away, otherwise the problem will only come back and often be worse in the long term.
   - Excessive use of avoidance strategies is often linked with several damaging effects, such as an increased susceptibility to mental health problems and stress-related illness.

   1 mark for each benefit (up to a maximum of 2 marks); 1 mark for each limitation (up to a maximum of 2 marks)

Chapter 2 Area of Study 2

Models to explain learning

Multiple-choice answers

1 D

Learning is any relatively permanent change in behaviour resulting from experience.

A is incorrect because it is too broad. Temporary behavioural changes, such as due to drugs or illness, does not fit the definition of learning. **B** is incorrect because learned behaviours are not temporary. **C** is incorrect because behavioural change is not due to experience, but when the person or animal is developmentally ready to perform the behaviour.

2 C

Classical conditioning is said to have taken place when a stimulus consistently produces a response even though it did not initially produce that response.

A pertains to observational learning. **B** would be indicative of a reflex. **D** pertains to operant conditioning.

3 B

Classical conditioning requires the development of an association between two stimuli, such that either one will trigger the response.

A is incorrect because it is the association formed between the neutral stimulus and the unconditioned stimulus which triggers the conditioned response. **C** is incorrect because the unconditioned response would occur due to the unconditioned stimulus, not the conditioned stimulus. **D** is incorrect because these are the same stimulus at different stages in the process.

4 C

In Pavlov's experiments, the food is the unconditioned stimulus, and the salivation is the unconditioned response.

A is incorrect because the aspects described pertain to the innate reflexive behaviour being studied, not the learned behaviour developed during the experiments. **B** is incorrect because the terms are reversed. **D** is incorrect because the conditioned response would be salivation in response to the bell.

5 A

In classical conditioning, learning is quickest when the neutral stimulus occurs slightly (about ½ second) before the unconditioned stimulus.

B is incorrect because the unconditioned stimulus would have already triggered the unconditioned response, making the neutral stimulus irrelevant. **C** is incorrect for similar reasons as above, although an association may develop over a longer period of repeated exposure. **D** is incorrect because it is possible that the person or animal would become habituated to the neutral stimulus and therefore ignore it.

6 B

An unconditioned stimulus is an innate stimulus that has the capacity to elicit a reflexive response.

A is incorrect because this refers to the conditioned response. **C** is incorrect because this refers to the unconditioned response. **D** is incorrect because this refers to the conditioned stimulus.

7 D

This would constitute a classically conditioned response. Loimata has associated the syringe (neutral stimulus [NS]) with the pain (unconditioned stimulus [UCS]) that led to her crying (unconditioned response [UCR]). This association triggered her crying (conditioned response [CR]) at the sight of the syringe (conditioned stimulus [CS]).

Despite being a reflexive response, **A** is incorrect because it is not innate and therefore not a reflex as such. **B** is incorrect because Loimata is responding to the stimulus, not a consequence. **C** is incorrect because spontaneous recovery occurs after a learned response has been extinguished. The fact that there was a period of time between each event does not mean that extinction had occurred.

8 C

The inflation of the balloons to a large size has become a conditioned stimulus.

A is incorrect because the noise from the bursting balloons was an unconditioned stimulus. **B** is incorrect because Tahir's behaviour of over-inflating the balloons led to the conditioned stimulus. **D** is incorrect because there is nothing in the scenario to indicate that this was the case.

9 C

As Aldo reacts to one stimulus but not to other similar stimuli, he is demonstrating stimulus discrimination.

A is incorrect because he did not react in the same way to other similar stimuli. **B** is incorrect because this is a term from operant conditioning referring to learned rewards – irrelevant to this scenario. **D** is incorrect because the learned behaviour is still persisting.

10 A

To extinguish a classically conditioned response, the conditioned stimulus is presented without the unconditioned stimulus in order to break the learned association between them.

B is incorrect because the unconditioned stimulus will still lead to the same response. **C** is incorrect because this describes the conditioning process. As **B** is incorrect, then by default **D** is also incorrect.

11 B

Spontaneous recovery involves the sudden recurrence of a conditioned response after extinction has occurred.

While Mila's response is to another dog, **A** is not the best choice, as the scenario focuses on the recurrence of her fear, not that it has generalised to other dogs. **C** is not appropriate because the scenario implies that extinction had occurred, not that the learned behaviour was just dormant. **D** is incorrect because the prompt focuses on the sudden anxiety after the period of extinction.

12 D

Operant conditioning focuses on the consequences of one's actions.

A is too broad and could also refer to classical conditioning. **B** is incorrect because it pertains to observational learning. **C** is incorrect because individuals may ignore instructions if the consequence for doing so is seen as more desirable than compliance.

13 D

The descriptions in the prompt are both describing how a desirable consequence may be applied; that is, reinforcement.

Positive reinforcement involves giving something good for the desired behaviour, whereas negative reinforcement involves taking away something bad if the desired behaviour is shown.

A and **C** are incorrect because punishment is not relevant to the prompt. **B** is incorrect because the terms are reversed.

14 C

Having your pain relieved by going to the dentist means that you are more likely to do the same thing next time you have a toothache.

A is incorrect because this is negative punishment (response cost). **B** is incorrect because this pertains to observational learning (vicarious punishment). **D** is incorrect because this is positive punishment.

15 A

Dan quickly learned to speak with an Australian accent in order to make the teasing stop. Getting a desirable outcome by removal of an aversive stimulus is negative reinforcement.

B is incorrect because the boys were not praising Dan for his Australian accent. **C** is incorrect because Dan was not having anything taken from him for speaking with an Australian accent. **D** is incorrect because Dan's speaking with an Australian accent had increased, not decreased. (If the prompt had stated 'Dan quickly learned to stop speaking with an English accent as a result of …', then the correct answer would have been **D**, as the teasing involved the addition of something unpleasant in response to the behaviour.)

16 A

The process of punishment requires the application of an undesirable consequence.

B is incorrect because punishment should decrease the response rate. **C** is incorrect because this relates to negative reinforcement. **D** is incorrect because this pertains to positive reinforcement.

17 B

As it would appear that Jake desires attention in any form, the intended punishment is perceived by him as a reinforcer.

A and **D** are inappropriate as there is nothing in the scenario to indicate that either of these is the case. **C** is incorrect because Jake's behaviour has not been inhibited.

18 B

For extinction to occur within operant conditioning, total cessation of reinforcement needs to occur.

A is incorrect because this only suppresses the behaviour, it does not eliminate it. **C** is incorrect because this is the process in classical conditioning. **D** is incorrect because it places total control within the individual and ignores the key aspect of this theory; that is, the effect of the consequences.

19 D

The process employed by the animal trainers to teach the dolphins is called shaping, which applies the principles of operant conditioning by reinforcing successive approximations to the desired behaviour.

A is incorrect because it implies that the dolphins are trying to figure out how to jump through the hoop, not the trainer trying to get them to do it. **B** is incorrect because the dolphins are not watching a demonstration of the desired behaviour in order to copy it. **C** is incorrect because this process pertains to the treatment of phobias, not animal training.

20 C

Observational learning entails watching others.

A is incorrect because the knowledge is gained indirectly; that is, through others, before any attempt at direct experience when trying to reproduce it. **B** is incorrect because this type of learning requires the learner to see the behaviour, not just hear about it. **D** is incorrect because the process involves thinking about the actions of others.

21 B

The order of the processes of observational learning outlined by Bandura are attention, retention, reproduction, motivation and reinforcement. The other options have incorrect arrangements of these concepts.

22 C

You would expect children who observed aggressive adults to engage in significantly more aggressive actions in subsequent play.

A is incorrect because Bandura focused on the observable effects on subsequent behaviour, not on possible emotional effects. **B** is incorrect because Bandura demonstrated observable changes in the children's patterns of play. **D** is incorrect because this is the opposite of Bandura's findings.

23 D

Vicarious classical conditioning involves people learning to fear an object or event by observing, then modelling, the reactions of others.

A is incorrect because the response described in the scenario was learned, not innate. **B** is incorrect because the scenario does not describe a fear of other bodies of water. **C** is incorrect because the learning has not extended upon prior learning.

24 A

The various TAC advertisements rely on vicarious operant conditioning to get their message across through different scenarios showing the harmful results of bad driver behaviour (vicarious punishment), fines and/or loss of licence (vicarious response cost), or the benefits of good behaviour (vicarious positive reinforcement).

B is ambiguous, but incorrect either way. The TAC advertisements do not need continual airplay and can be very effective if only seen a small number of times. If **B** was talking about the actual driving behaviour, then reinforcement would be delivered directly, not through the media. **C** may apply in advertising products, but it is incorrect in this context, as TAC advertisements emphasise the consequences (good or bad) resulting from certain driver behaviours. **D** is incorrect because viewers do not need a period of time to think about the content before realising its meaning.

Short-answer sample responses

25 a Learning is any relatively permanent change in behaviour resulting from experience. (1 mark)

b Reflex actions are innate, simple, automatic/involuntary responses that do not require prior experience; for example, the spinal reflex, or a person blinking when a puff of air is blown into the eye.

1 mark for an accurate description of a reflex; 1 mark for providing an appropriate example

26 a Any involuntary, automatic response, over which the individual cannot exert any control, can be classically conditioned. (1 mark) Such responses would include reflexes, emotional responses or other responses evoked by specific stimuli. (1 mark)

b A reflex would be described as an unconditioned stimulus triggering an unconditioned response. (1 mark)

c A conditioned stimulus is a previously neutral stimulus that, through association with the unconditioned stimulus (1 mark) has acquired the capacity to provoke a conditioned response (1 mark).

27 a Pair the unwanted *neutral stimulus* (the cigarette) with an *unconditioned stimulus* (such as an electric shock) that reflexively produces an undesirable *unconditioned response* (in this case, pain). (1 mark) After several pairings, the *conditioned stimulus* (the cigarette) will produce an aversive *conditioned response*, even when it is not paired with the *unconditioned stimulus* (the electric shock). (1 mark)

Many students lost marks in this exam question because they did not use the language of classical conditioning as instructed in the question prompt; that is, incorporate the terms italicised above.

b Any one of the following (1 mark):
- Failure to maintain the behaviour outside the clinical situation
- Overgeneralisation (not just giving up drinking alcohol, but develop aversion to other forms of drinking)
- Ethical considerations concerning causing the patient physiological or psychological harm
- Extinction of the conditioned response without repeated administrations of the pairing of the UCS + CS

28 a Operant conditioning involves learning in which the consequences of a behaviour determine the probability of its recurrence. (1 mark) In operant conditioning, an individual will increase or decrease the frequency of their behaviour depending on the consequences of their actions. (1 mark)

b According to operant conditioning, responses that are followed by a satisfying result would occur with greater frequency over time (1 mark), whereas responses that were followed by an undesirable consequence would occur less frequently over time (1 mark).

29 a Reinforcement refers to any consequence that strengthens or increases the probability of a response recurring. (1 mark)

b Punishment refers to any aversive or unpleasant consequence that weakens or decreases the probability of a response recurring. (1 mark)

30 a 'Positive' refers to the 'addition' of something as a consequence (1 mark), whereas 'negative' refers to the 'removal' of something as a consequence (1 mark).

> Common errors occur due to the association of 'positive' with 'good' and 'negative' with something 'bad', especially in the case of negative reinforcement, where you achieve a good result by taking away something undesirable.

b In negative reinforcement, you are removing an unpleasant or aversive consequence (or threat of that consequence) in order to increase the frequency of a certain behaviour (1 mark) (e.g. cleaning your room means that your parents will stop nagging you to do so, increasing the likelihood that you will keep it tidy; or not speeding while driving means that you won't get a fine and increases the likelihood that you will drive safely) (1 mark).

Punishment is any aversive or unpleasant consequence that weakens or reduces the incidence of a behaviour (1 mark) (e.g. you have to do community service for certain misdemeanours, or you have to pay a fine and lose points from your licence for speeding) (1 mark).

31 a The baby's behaviour crying each night was maintained by positive reinforcement (1 mark) as it led to a desirable outcome of being picked up and comforted by her father and rocked until she went back to sleep (1 mark).

b Mike's behaviour of going to the baby's room as soon as she started to sob was maintained by negative reinforcement (1 mark) as it led to a desirable outcome by stopping the baby's crying (1 mark).

> While getting some sleep is also a desirable consequence that might be considered a positive reinforcer, this would only be possible after the removal of the unpleasant stimulus, making it an after-effect rather than a direct consequence driving the behaviour.

32 a The learning principle was negative reinforcement. (1 mark) If the desired behaviour (homework) occurs, then the unpleasant stimulus (her father's pestering) will stop. (1 mark) It is the termination of the unpleasant stimulus that acts as the reinforcement to motivate Najida to perform the desired behaviour (do her homework). (1 mark)

b The token economy acts as an incentive by providing positive reinforcement for Najida doing her homework. (1 mark) Whenever Najida completes her homework without her father pestering her, he would reward her with a token, such as putting points or ticks on a chart. (1 mark) An agreement is made whereby a certain number of tokens can be exchanged for a desired reward (e.g. certain privileges). (1 mark)

    c The learning principle was negative punishment (or response cost). (1 mark) The act of taking away a desired activity as a consequence for an undesirable behaviour (i.e. not completing homework) is a punishment that may motivate Najida to do her homework in order to avoid this punishment. (1 mark)

33 a Any one of the following (1 mark):
- Polly hatching/breaking through the eggshell (due to maturation)
- Polly screeching (due to maturation)

  b When Polly hears the aviary door open (*conditioned stimulus*), she gets excited and screeches (*conditioned response*). (1 mark)

  c Marco could use shaping (or operant conditioning, or method of successive approximations). (1 mark) Marco could reward Polly with a food treat whenever she makes an appropriate sound. (1 mark) Then, over time, he would reward only sounds that get progressively closer to the desired sound until a word is said. (1 mark)

> It was essential that responses related to the scenario. Generic descriptions of an appropriate learning technique would only be awarded a maximum of 1 mark.

34 a
- When two things occur together, we associate one with the other so that when one appears we expect the other to follow and react accordingly. Both forms of conditioning are based on this process. (1 mark)
- For classical conditioning, an association is formed between two stimuli such that both will trigger the same response. (1 mark)
- In operant conditioning, an association is formed between a behaviour and the events that follow it such that the latter will be seen as occurring as a result or a consequence of that behaviour. (1 mark)

  b **i** Classical conditioning involves reflexive, involuntary behaviours (1 mark), whereas operant conditioning and observational learning predominantly involve non-reflexive, voluntary or intentional behaviours (1 mark).

    ii In classical conditioning, the learner is passive, as the response is caused by the unconditioned stimulus or, later, by the conditioned stimulus (1 mark), whereas in operant conditioning and observational learning, the learner is active as the individual chooses to respond in a particular manner in order to lead to reinforcement or avoid punishment (1 mark).

    iii In classical conditioning, reinforcement occurs *before* the response (characterised by *antecedent* events) (1 mark), whereas in operant conditioning reinforcement occurs *after* the response as a *consequence* of the behaviour (1 mark). In observational learning, reinforcement also occurs *after* the response as a *consequence* of the behaviour, but learning is not necessarily immediately evident, and reproduction of the observed behaviour could be delayed until desired/required. (1 mark)

35 a Bandura's theory is a form of operant conditioning because:
- according to social learning theory, we learn through observing the consequences of others' behaviours, whether they are reinforced or punished (1 mark)
- learning in which behaviour becomes controlled by its consequences is called operant conditioning. (1 mark)

  b Any two of the following (2 marks):
- The consequence that the model incurred as a result of their behaviour
- The amount of attention that the observer paid to the model
- The observer's ability to remember what the model did (retention)

- The observer's ability (or belief in their ability) to copy the modelled behaviour (reproduction)
- The amount of motivation that the observer has to repeat a task
- The characteristics of the model, such as similarity to the observer, attractiveness and trustworthiness
- The observer's admiration for the model, in terms of their status, expertise or power

c
- Attention: Jyotika intently watches her mother demonstrate the procedure a few times. (1 mark)
- Retention: Jyotika stores what she has seen in order to copy it later. (1 mark)
- Reproduction: Jyotika copies what she has seen her mother do by performing long-stitch onto the tapestry. (1 mark)
- Motivation: Jyotika continues to attempt her long stitch in order to get her mother's approval. (1 mark)
- Reinforcement: Jyotika shows what she has done to her mother who then praises her for doing a good job. (1 mark)

36 a Type of conditioning: The advertisement uses classical conditioning principles. (1 mark)

Reason 1: The drink (*neutral stimulus*) is repeatedly paired with images of people having fun (*unconditioned stimulus*) that promote a positive emotional response (*unconditioned response*). (1 mark)

Reason 2: With repetition, the drink becomes a *conditioned stimulus* that can produce positive feelings on its own (*conditioned response*). (1 mark)

b Process: Attention

Justification: If it is associated with someone famous, the viewer is more likely to actively focus on the drink and encode information about it.

Process: Retention

Justification: Because it is associated in memory with someone they admire, viewers are more likely to remember the product and the pleasant feelings connected to it.

Process: Motivation

Justification: The learner's need to try the drink is likely to be increased due to the association between the celebrity and the drink.

Process: Reinforcement

Justification: The learner may feel pleased to be able to emulate the celebrity by using the product (positive reinforcement).

> This question was marked as two separate 2-mark questions: 1 mark for identifying a correct process of observational learning (× 2) and 1 mark for a valid justification of the relevance of that process within the scenario (× 2). Reproduction was not considered to be relevant to this scenario.

The psychobiological process of memory

Multiple-choice answers

1 A

> Encoding (**B**), storage and retrieval (**D**) are considered to be the three basic processes of memory.
> **A**, attention, is required in order for information to pass into our short-term memory for processing but is not itself considered a memory process. **C** is an alternative term for storage.

2 C

The function of sensory memory is to briefly save our sensory impressions so that a slight overlap occurs, thereby enabling us to perceive our environment in an uninterrupted fashion rather than as a series of disjointed images and sounds.

While sensory memory is the initial stage of memory where external stimuli are registered, the information is only held temporarily by the sense organs unless it proceeds into short-term memory (hence **A** is incorrect). **B** is incorrect because sensory memory is immense in its capacity, restricted only by the receptive field of the sense involved. Although the course does not emphasise tactile (touch) memory, it is a type of sensory memory in addition to iconic (visual) memory and echoic (auditory) memory (therefore **D** is incorrect).

3 D

Through selective attention, an individual is able to focus on the relevant information so that it can be processed in their short-term memory.

A is incorrect because this expands the capacity of short-term memory. **B** is incorrect because this occurs once the information has entered short-term memory. **C** is incorrect because this is the process whereby relevant information becomes stable in long-term memory.

4 A

Short-term memory stores a limited amount of encoded information while it is required for further manipulation and processing.

While chunking can expand the capacity of short-term memory, **B** is incorrect because it is not a requirement in every case. **C** is incorrect because information can also come from the sensory register. **D** is incorrect because short-term memory is limited in capacity, so cannot process all of the sensory data.

5 C

Miller's research article 'The Magic Number 7 plus or minus 2' was referring to the storage capacity of short-term memory.

According to this research, the average number of 'bits' of information that can usually be processed within short-term memory is seven. **B** and **D** refer to the outer limits of spread expected from Miller's research. **A** is incorrect because this is well below the average.

6 B

If prevented from being able to rehearse the information, the duration of short-term memory is about 18 to 20 seconds, after which participants' ability to recall the words would rapidly decline.

A implies the decline of echoic memory which is irrelevant to the scenario. **C** is referring to the capacity of short-term memory and would be true if there was no interference involved. **D** is incorrect because the difference in stimuli is irrelevant as interference does occur to prevent rehearsal.

7 A

Elaborative rehearsal establishes connections with declarative information already in long-term memory, thereby making it easier to remember.

B is incorrect because this expands short-term memory but does not guarantee that the information will proceed into long-term memory. **C** is incorrect because this is the process described in the first part of the prompt. **D** is incorrect because this implies the development of a motor skill, not the processing of information.

8 A

Declarative memory incorporates autobiographical information (episodic memory) and factual knowledge (semantic memory).

B and **D** are incorrect because they only name one aspect of the prompt. **C** is incorrect because this relates to motor skills.

The key word in the prompt is '**and**', indicating that the name required is an umbrella term incorporating a number of concepts. Misreading by focusing only on the first phrase or on the latter part of the question would lead to an error in this case.

9 D

D pertains to short-term memory, not long-term memory.

The other options do describe aspects of long-term memory, which is contrary to the specific wording of the question.

10 C

The scenario contains elements that could apply to any of the alternatives given, so it is important to read the prompt carefully.

Yuki's retained ability to ride a bike would be stored in her procedural memory (**C**).

A is incorrect because this is involved when she is articulating her memories. **B** is incorrect because this type of memory was involved when describing specific events she had experienced. **D** is incorrect because this was used to recall factual information about the steps she took in her learning.

11 B

Consolidation refers to the period of time necessary for a lasting long-term memory to develop.

A is incorrect because consolidation is the process required for stable long-term memory, not short-term memory. **C** is incorrect because this would be an aspect of encoding and elaborative rehearsal. **D** is incorrect because electroconvulsive therapy (ECT) has been shown to disrupt consolidation, not speed it up.

12 D

Procedural memory is a type of implicit memory of skills that are processed and stored within the cerebellum.

The fact that the amygdala was mentioned in two options does not mean that the correct answer is most likely one of these. **B** is incorrect because the amygdala focuses on the consolidation of implicit memories involving emotion, not skills as required by the prompt. While the amygdala may process emotional and state-dependent aspects, **C** is incorrect because consolidation of explicit episodic memory of life events requires the action of the hippocampus. While the hippocampus is actively involved in the storage and consolidation of explicit (or declarative) memory, this type of memory deals with the retention of information, not skills. (Therefore, **A** is not correct.)

13 C

The hippocampus is **not** clearly implicated in the formation and retrieval of implicit procedural memories (which occur due to the actions of the cerebellum and basal ganglia).

The other options do describe characteristics of the hippocampus, which is contrary to the specific wording of the question.

14 B

Alzheimer's disease begins slowly, and at first the only symptom may be mild forgetfulness. People in the early stages of Alzheimer's disease may have trouble remembering recent events, activities or the names of familiar people or things. The other options may occur in more advanced stages of the disease.

15 C

Alzheimer's disease is the most common form of dementia.

A is incorrect because it is not a normal part of ageing. **B** is incorrect because the exact cause of Alzheimer's disease is still unknown. **D** is incorrect because there are definite physiological markers that can only be found through autopsy of the brain.

16 A

The defining characteristic of Alzheimer's disease is deficits in all aspects of memory.

B is incorrect because new long-term memories cannot be made without proper processing in short-term memory. **C** is indicative of the early stage of the disease, but the prompt points to the progressive loss of long-term memory in the later stages. **D** is incorrect because short-term memory is disrupted, even from the early stages of the disease.

17 D

Aphantasia is the inability to visualise images voluntarily in one's head.

A and **C** are incorrect because they apply to anterograde amnesia. **B** is incorrect because they can describe factual information, they just can't picture it with their mind's eye.

18 C

Functional MRI scans of individuals with aphantasia have found that when they tried to visualise imagery, there was a significant reduction in activation patterns within the occipital lobes, while activity in the frontal regions responsible for decision-making and error prediction was significantly increased compared to controls.

A and **B** incorrectly describe a reduction in activity in the frontal regions. **B** is also incorrect for describing increased activation in the occipital lobes. **D** is incorrect because there were significant differences recorded.

19 B

The words SCUBA and LASER are examples of acronyms as they combine the initials of several concepts to create a pronounceable word.

A is incorrect because this method creates sentences, not words. **C** is incorrect because using location as a cue for memory is not involved in this case. **D** is incorrect because imagery is not involved in these examples.

20 C

Alexi used the method of loci to attach the required concepts to a sequence of locations in his house; Bethany used acrostics to create sentences where each word began with the letters of the required concepts in their correct order.

A is incorrect because Alexi did not use acrostics (as explained above) and Bethany did not apply Indigenous Australian methods using Songlines. **B** and **D** are incorrect because neither Alexi nor Bethany created pronounceable acronyms using the initials of the required concepts.

Short-answer sample responses

21 a Memory is an active, information processing system that receives, organises, stores and recovers information. (1 mark)

b Encoding transforms incoming information into a usable form so that it can go into memory. (1 mark) Storage (or retention) refers to the process of maintaining information in our memory. (1 mark) Retrieval is the process of recovering information that has been stored. (1 mark)

c For information to be saved, it must first be encoded; that is, converted into a form that can be incorporated into the networks in our memory. (1 mark) The process of storage is required to retain this information for later use. (1 mark) Retrieval is also an essential process as, without it, the individual would not be aware of the stored information and any such material would be useless. (1 mark)

22 Sensory memory is the initial stage of memory:
- with immense capacity
- but only capable of storing information for very brief periods (a fraction to a few seconds).

Short-term memory:
- holds a limited amount of information (approximately seven plus or minus two items)
- for 18–20 seconds.

Long-term memory:
- theoretically has an unlimited capacity
- is relatively permanent, storing information for extensive periods of time (up to a lifetime).

1 mark each for identifying the stages of memory in the correct sequence (× 3); another mark each for describing their capacity (× 3); a further mark each for correctly describing their duration (× 3)

23 a The key function of sensory memory is to briefly save our sensory impressions so that a slight overlap occurs, thereby enabling us to perceive our environment in an uninterrupted fashion rather than as a series of disjointed images and sounds. (1 mark)

b Any two of the following (2 marks):

Iconic memory:

- is a type of sensory memory for visual information
- stores information in a raw, literal, unencoded form
- retains visual images for about 0.3 seconds before it starts to rapidly fade away
- enables us to clearly distinguish environmental stimuli rather than visual images overlapping or blurring together
- allows us to perceive smooth motion rather than discrete, disjointed images (as in a movie).

c Echoic memory is the sensory memory which retains auditory information (1 mark) for 3–4 seconds (1 mark) so that we can process sounds and interpret their meaning.

d Information held in sensory memory cannot be rehearsed to prolong its duration because we have no voluntary control over its rapid decay. (1 mark)

24 a Information for short-term memory comes from the material we pay attention to within our sensory register (1 mark) and/or information retrieved from our long-term memory (1 mark).

b An advantage of short-term memory having a small storage capacity is that we are better able to focus on relevant aspects of the material held there. (1 mark) By dealing with smaller amounts of information, it is easier to search through it while trying to process, manipulate and/or encode the data being held in short-term memory. (1 mark)

c Chunking involves grouping items into larger units to extend/expand the capacity of short-term memory. (1 mark)

25 a Because short-term memory fades after about 20 seconds (1 mark), the process of rehearsal serves to renew the information via some form of mental repetition or processing (1 mark), so that it can be retained in short-term memory for a longer period of time (1 mark).

b Maintenance rehearsal (1 mark) is the form of rehearsal that uses the conscious recitation of information so that it can be retained in short-term memory for a brief period (1 mark). The material remains in short-term/working memory for only a short while and, as it is in a meaningless, rote form, is not effectively encoded into long-term memory. (1 mark)

c Elaboration (or elaborative rehearsal) helps us to store information in our long-term memory as this process involves deeper processing of new information by making it more meaningful as a result of linking it to material already stored in our long-term memory. (1 mark) By actively thinking about the information to process it in this way, we are maintaining it within our short-term memory for a longer period of time, thereby giving it more time to proceed into our long-term memory. (1 mark) A range of examples would be acceptable here, including mnemonic techniques, mental imagery or other instances that link material in new, novel or bizarre ways. (1 mark)

26
- Explicit (or declarative) memory contains information that can be consciously brought to mind and verbally communicated to others, including knowledge of specific facts or events. This type of memory consists of two subcategories: semantic memory and episodic memory. (1 mark)
- Semantic memory stores specialised factual information, including general knowledge, academic concepts of the variety learned at school, as well as the meaning of words. (1 mark)
- Episodic memory represents autobiographical details about an individual's personal experiences, including times, places and proceedings of specific life events. (1 mark)
- Implicit (or non-declarative) memory stores automatic responses that have been learned through experience. This type of memory includes procedural memory and conditioned/emotional responses. (1 mark)

- Procedural memory involves thought processes and skills about how to perform a task that enable an individual to carry out a course of action; for example, riding a bike or tying one's shoelaces. (1 mark)
- Conditioned/emotional responses are reactions that occur without conscious effort, but which are not innate, and so must have been stored after some prior experience. (1 mark)

27 a Any one of the following (1 mark):

Research suggests that the hippocampus:
- is critical to the formation and consolidation of memories
- plays a part in deciding if information received by the senses is worth remembering
- maps and organises memories before directing them to other sections of the brain
- is central to recalling spatial relationships in the world about us.

b Any one of the following (1 mark):

The amygdala:
- is associated with processing and consolidating implicit memories for emotional events
- modulates the hippocampal formation during times of arousal to improve the encoding and consolidation of subsequent long-term memory
- plays a part in determining which memories are encoded and stored
- is associated with processing and consolidating implicit conditioned responses, such as fear.

c Any one of the following (1 mark):

The neocortex (cerebral cortex):
- processes information in order to encode it for storage
- stores long-term explicit, declarative memories.

d The basal ganglia is involved in learning skills that result in implicit procedural memory or habit formation. (1 mark)

e Any one of the following (1 mark):

The cerebellum is:
- involved in the formation, storage and retrieval of implicit procedural memories
- responsible for the enactive (muscle) memory for fine motor skills.

28 a The early stages of Alzheimer's disease are characterised by difficulty in remembering episodic memories. (1 mark) There can also be problems in remembering new/recent semantic information. (1 mark)

b Any two of the following (2 marks):
- Senile plaques
- Accumulation of fatty deposits around sticky molecules of beta-amyloid protein within the brain
- Neurofibrillary tangles
- Deterioration of the hippocampus
- Atrophy (wasting away) of brain tissue
- General brain shrinkage from cell death
- Abnormal clusters of dead and dying nerve cells
- Destruction of neurons involved in production of some neurotransmitters, especially acetylcholine

29 a Semantic memories would provide the factual information (when, where, who with, etc.) (1 mark), while the episodic memory would store perceptions of the event from your individual viewpoint, including how you felt at the time (1 mark).

b Someone with aphantasia would probably describe their autobiographical memories in semantic terms (1 mark) but they would not be able to form a mental image to provide the episodic elements of the events of that day (1 mark).

30 Method of loci involves visualising an image linking each item to be remembered with locations in a familiar route or sequence of places that can be easily recalled from the long-term memory (such as rooms in your home or landmarks on your way to school). (1 mark) By retracing the route in your head and examining the images, you can recall the items so that you can reconstruct the concepts in order. (1 mark) The technique is best for recalling lists of items (in a particular order). (1 mark)

31 a Acrostic: The example needed to be a coherent sentence, with each word beginning with the first letter of each place to be visited, maintaining the order of places. For example, 'Peter Often Sees Helicopters' (or, if post office was treated as two words: 'Pirates Only Own Silver Hooks'.) (1 mark)

Acronym: The example needed to use the first letter of each place on her list to form a pronounceable word; in this case POSH (or POOSH). (1 mark)

b Acrostics and acronyms are mnemonic devices (memory strategies/techniques) that (1 mark) provide cues/hints to facilitate retrieval of the information in the required order/sequence (1 mark) through elaboration and organisation of the material in long-term memory (1 mark).

32 Songlines are multimodal because they are expressed through song, dance and symbols used in art and cultural artefacts (1 mark). Songlines are situated because they are linked to significant places on Country that are associated with specific knowledge, ceremonies and trade routes (1 mark). The Songlines are said to be patterned on Country because they form a network of routes across Country (1 mark) that form links between places of significance and the knowledge associated with these places (1 mark).

UNIT 4

Chapter 3 Area of Study 1

The demand for sleep

Multiple-choice answers

1 B

'Consciousness' is defined as the subject's awareness of objects and events in the external world and their own existence and activities. **A** and **D** only address part of this definition. **C** is too vague, especially as the various levels or states of consciousness have different types of physiological activity.

2 C

As consciousness exists along a continuum that blends from one state to another, it has been described as an ever-changing stream of awareness.

A is incorrect because the concepts of mind and body are distinct from one another, with debate over where consciousness would exist. **B** is incorrect because consciousness is not a stable, fixed state. **D** is incorrect because individuals are cognisant of their thoughts and awareness in the higher levels of the continuum, especially in the normal waking state.

3 C

All other characteristics are consistent with being in a state of ordinary or normal waking consciousness except the disruption of continuity of our memory over time.

The prompt for this question could also have been: 'Identify which of the following characteristic is consistent with being in an altered state of consciousness'.

4 D

The flow of thoughts described for Rebekkah conforms to the changing thought patterns within our normal waking consciousness.

A is incorrect because memories and planning are not fantasies. **B** is incorrect because Rebekkah is aware of and directing her thoughts towards each topic. **C** is incorrect because Rebekkah is not intently focused on one thing.

5 D

Electrodes for the EEG go on the person's scalp, while those for the EOG are near the eyes and those for the EMG tend to be on the jaw muscle.

EMG electrodes in region 1 would not detect any activity from muscles, making **A** incorrect. The EOG electrodes need to be near the eyes, making **A**, **B** and **C** incorrect. To measure brain activity, EEG electrodes need to be in region 1, making **A** and **B** incorrect.

6 C

Sleep diaries and video monitoring are qualitative measures, making **C** correct and **A** and **D** incorrect. Devices such as the EEG, EMG and EOG provide quantitative measurements, confirming that **C** is correct and that the other options are incorrect.

7 D

The EEG provides information about brain activity, the EMG provides information about muscle tone and activity, and the EOG provides information about patterns of eye movement.

A is incorrect because the information for the EMG and EOG are reversed. **B** is incorrect because none of the information matched any of the devices. **C** is incorrect because the information for the EEG and EMG are reversed.

8 B

Beta brain waves are present during an alert state of normal waking consciousness.

A is incorrect because alpha waves are evident in a state of relaxation, meditation or light sleep. **C** is incorrect because delta waves occur in deep sleep. **D** is incorrect because theta waves occur during the sleep stages.

9 A

Brain wave recordings indicate that during sleep, the brain changes in its pattern of activity.

B and **C** are incorrect because the brain is still active and functioning while we are asleep. **D** is incorrect because this only pertains to REM sleep, not the whole sleep cycle.

10 D

The correct sequence of brain waves as we descend into deep sleep is beta, alpha, theta, delta.

A is incorrect because this is merely the beginning of the Greek alphabet. **B** is incorrect because it presents the letters in their order within the Greek alphabet. **C** is incorrect because alpha and beta should be swapped.

11 A

During REM sleep, EEG and EOG recordings show high activity whereas the EMG shows low activity.

B is incorrect because both of these measures should show high activity. **C** is incorrect because both levels of activity are opposite to what would be the case. **D** is incorrect because the EMG would show low activity.

12 D

A typical sleep cycle for an adult from early in the night would take them from being awake to progressing through all the stages of NREM, then back up again before the first occurrence of REM.

A is incorrect because we do not go instantly into NREM stage 3 and work our way back up to REM. **B** is incorrect because hypnograms clearly indicate that brief periods of intermediary NREM stages are experienced between NREM stage 3 and the REM stage. **C** is incorrect because we experience the hypnogogic state, not REM, as we go from being awake into sleep. The NREM stages are also reversed in **C**.

13 D

Hypnograms show that periods of REM sleep become progressively longer and occur more frequently within the progression across a normal night's sleep for a young adult.

A is incorrect because NREM sleep gets progressively shallower in each subsequent cycle throughout the night, with NREM 3 only occurring in the early cycles. **B** is incorrect because young adults spend approximately 80% in NREM and 20% in REM. **C** is incorrect because each sleep cycle lasts approximately 90 minutes.

14 D

NREM sleep helps to restore bodily processes and repair tissue damage.

A is incorrect because NREM sleep becomes shallower throughout the night. **B** is incorrect because there is more NREM sleep in the first half of the sleep cycle. **C** is incorrect because NREM sleep comprises about 80% of the time spent asleep.

15 C

C is not a function of REM sleep as the brain activity observed during this stage is similar to that recorded in subjects who are wide awake.

All the other options have been proposed as functions of REM sleep.

16 C

The restorative theory of sleep would explain why longer periods of sleep (especially NREM stage 3) occur after heavy physical exercise.

A is incorrect because NREM sleep helps to restore bodily processes and tissue damage, while REM sleep restores brain processes. **B** and **D** are incorrect because children need more sleep, including higher levels of REM, than adults.

17 A

Circadian rhythms are biological cycles that are approximately 24 hours in duration.

B and **D** are incorrect because these apply to longer infradian rhythms. **C** is incorrect because this is an ultradian rhythm, such as the cycles during sleep.

18 C

Results of the cave studies found that the human participants adapted their physiological processes to a regular 25-hour cycle.

A is incorrect because there was a change in the length of the circadian cycle. **B** and **D** are incorrect because there was no significant change to the amount of sleep compared to baseline norms.

19 B

The suprachiasmatic nucleus (SCN) responds to differing intensities of light and subsequently regulates the activity of the pineal gland in the secretion of melatonin to control the sleep–wake cycle.

A is incorrect because the pineal gland does not itself respond to light but secretes melatonin according to messages from the SCN. **C** is incorrect because the hypothalamus has different sections controlling functions to do with various types or arousal, but it is the SCN that responds to light. **D** is incorrect because this group of structures is involved in the voluntary control of movement and formation of procedural memory.

20 D

The hormone melatonin, secreted by the pineal gland, seems to be a key factor in regulating the sleep–wake cycle.

A is too broad and incorporates other states that are not regulated by the melatonin cycle. **B** is incorrect because this applies to adrenaline, not melatonin. Even though body temperature follows a daily cycle, **C** is incorrect because it is not regulated by melatonin.

21 B

As people age across the life span, the percentage of time spent in REM sleep decreases.

A is incorrect because REM is not always preceded by the three stages of NREM; as sleep progresses throughout the night, NREM gets shallower, with stage 3 sleep being omitted before later episodes of REM. While REM does increase in the second half of the night, **C** is incorrect because REM periods get longer, not shorter. **D** is not consistent with theories trying to explain the differing purposes of REM and NREM sleep.

22 D

Out of the options given, **D** gives the best set of figures for the three age groups.

A is incorrect because the hours for the infant are too low. **B** is incorrect because the hours slept are lower than expected, especially for the adolescent, and the percentages of time in NREM and REM are reversed in each case. **C** is incorrect because the first set of figures is indicative of a newborn baby, not an infant. Further to this, in both **A** and **C**, the hours for the adolescent and old age are swapped and should have 80% in NREM and 20% in REM sleep for adolescents, and approximately 85% NREM and 15% REM for the elderly.

23 C

A, **B** and **D** are incorrect because infants typically spend significantly more time in REM sleep than is indicated in either of the hypnograms.

Short-answer sample responses

24 a 'Consciousness' is defined as the awareness of one's own existence and mental activities (including thoughts, sensations and feelings) (1 mark) and of objects and events in the external world (1 mark).

b The term 'state of consciousness' refers to the levels within a continuum of awareness (1 mark) that an individual has of both their internal processes and external stimuli (1 mark).

c Any three of the following (3 marks):
- Awareness of thoughts, feelings and sensations
- Thinking is clear, organised and meaningful
- Attention can be focused on the task at hand
- Information can be retrieved from memory
- Awareness of an individual's identity
- Conscious of time limitations
- Not easily influenced by suggestions

25 a While in a state of normal waking consciousness, we are aware of the variety of sensory inputs from the outside world along with our thoughts, feelings and reactions to these inputs. (1 mark) Because we spend most of our life in this state, most people assume that this is our experience of 'reality', and it is this state that provides the baseline by which we judge all other states of consciousness. (1 mark)

b
- By comparing their speed and accuracy to baseline measures or expected norms for performance while in the normal waking consciousness, the measurement of performance on a variety of cognitive tasks can be used as an indicator of an individual's level of consciousness. (1 mark)
- While speed on various cognitive tasks may be enhanced when in a highly aroused state, including when using strong stimulants, accuracy is not assured as simple errors can occur. (1 mark)
- In cases of lowered arousal, speed would be slower and inaccuracies would occur, especially in the performance of simple tasks. Deficits in cognitive performance would become increasingly more obvious as the individual progresses down the consciousness continuum. (1 mark)

26 a Any three of the following (3 marks):
- Difficulty maintaining attention
- Cognitive distortions, including illogical, disorganised or disjointed thinking, which could affect the ability to form accurate memories and even cause memory lapses/blackouts
- Perceptual distortions, including changes to pain thresholds, experiencing sensory stimulation more vividly or losing the sense of where the self ends and the environment begins (such as the feeling of sinking into the floor when meditating)
- A disturbed time-sense (time may appear to slow down or speed up)
- Changes in emotional feeling (may feel more emotional or more subdued than normal)
- Changes in self-control, including difficulty controlling or coordinating movement, a loss of usual inhibitions and an openness to the influence of suggestions

b Naturally occurring states of consciousness are those that occur as part of the natural daily cycle, such as sleep and dreaming (1 mark), whereas those that have been induced do not naturally occur, but rather have been purposefully generated or brought about by other factors, such as drugs or external influences (1 mark).

c Any three of the following (3 marks):
- Thought patterns are disorganised
- Distorted perception of time
- Lack of control of movements
- Sensations and perceptions are dulled
- Awareness of external environment is reduced
- Memory is impaired

27 a Physiological responses give a better indication of differing states of consciousness, as these are objective measures that also provide information about bodily functions in those states. (1 mark) Self-report methods provide subjective data which is open to bias or inaccuracy. (1 mark)

b Any two of the following (2 marks):
- Individuals can be dishonest.
- Individuals may be unable to describe their experiences in words.
- Individuals may neglect, intentionally or unintentionally, to provide crucial information to the researcher.

c Video monitoring uses infrared cameras to monitor a person's movements during sleep. (1 mark) Video monitoring of the eyes would detect an increase in the rate of eye movements under eyelids (i.e. rapid eye movements). (1 mark)

28 a Amplitude (1 mark)
 b Frequency (1 mark)
 c REM sleep (1 mark)
 d NREM stage 3 (1 mark)

29 a An electroencephalograph (EEG) detects, amplifies and records electrical activity of the brain (in the form of brain waves). (1 mark)

b An electromyograph (EMG) detects, amplifies and records electrical activity in the muscles. (1 mark)

c The electro-oculograph (EOG) detects, amplifies and records electrical activity of the muscles that control eye movement. (1 mark)

> Responses that state 'electrical activity in the eyes' or 'the activity of the eye muscles' would not receive any marks.

30 a NREM stage 2 (1 mark)

b The unique brain wave features that are characteristic of stage 2 sleep are:
- sleep spindles – brief bursts of higher-frequency brain waves
- K complexes – single sharp spikes (rise then fall) in amplitude (and lower frequency). (2 marks)

> 1 mark for identifying each feature; 1 mark for correctly describing each feature (theta waves are not 'unique brain wave features' characteristic of stage 2 sleep, as they also occur in stage 1 and as we transition in and out of stage 3)

31

Sleep stage	EEG recording	EMG recording	EOG recording
REM	Fast, irregular, unsynchronised, low-amplitude beta-like waves (1 mark) indicating high levels of electrical activity in the brain (1 mark)	Extremely low levels of electrical activity in the muscles (1 mark) indicating little or no muscle movement (1 mark)	High levels of electrical activity in the muscles that control eye movement (high-frequency, low-amplitude waves) (1 mark) indicating rapid eye movement (1 mark)
NREM stage 3	Slow, regular, synchronised, high-amplitude delta waves (1 mark) indicating low levels of electrical activity in the brain (1 mark)	Generally low levels of electrical activity in the muscles (1 mark) indicating limited (and irregular) muscle movement (1 mark)	Very low (or no) electrical activity in the muscles that control eye movement (few waves/little electrical activity in the muscles around the eye) (1 mark) indicating little or no eye movement (1 mark)

> Descriptions of the recordings need to refer to the type or level of electrical activity, especially as the latter part of the prompt points to the type of behaviour that can be inferred from the measurements.

32 a An adult sleeps 7–8 hours per night, progressing through cycles of about 1.5 hours. (1 mark)
- Periods of deep sleep (NREM stage 3) occur earlier in the night and NREM cycles get progressively shallower as the night goes on. (1 mark)
- Periods of REM sleep occur, on average, every 90 minutes, with each session of REM progressively increasing in duration as the night goes on, starting at 10 minutes and reaching 30+ minutes in the morning. (1 mark)

b Any two of the following (2 marks):
- Erratic, high-frequency, low-amplitude brain activity reminiscent of beta waves
- Increased electrical activity in eye muscles (i.e. rapid, erratic eye movements)
- Little or no electrical activity in body muscles (i.e. muscles relaxed to the point of paralysis, atonia)
- Other autonomic levels (such as breathing, heart rate) increased

c NREM comprises about 80% of a typical adult's nightly sleep cycle (1 mark), with more NREM sleep present during the first half of the night than the second half (1 mark).

33 a Any two of the following (2 marks):
- The duration of REM cycles becomes longer as the night progresses, whereas NREM cycles become shorter.
- Deep sleep (NREM stage 3) only occurs in the first couple of cycles, whereas REM cycles occur throughout the night.
- Across the total sleep period, more time overall is spent in NREM compared to REM.

- NREM sleep has three stages whereas REM has only one.

b An adolescent is likely to have more REM cycles than an elderly person. (1 mark) An adolescent will have a different pattern of NREM sleep with more NREM stage 3 than an elderly person, who would have little or no NREM stage 3, but more NREM stages 1 to 2. (1 mark)

34 a
- As individuals age, they require less total sleep per 24-hour period. (1 mark)
- Babies sleep 18–20 hours, toddlers 13–14 hours, adolescents 9–10 hours, adults 7–8 hours and elderly people 6 hours with less deep sleep. (1 mark)
- The percentage of total sleep time spent in REM sleep also lessens as we get older. (1 mark)
- A newborn spends approximately 50% of total sleep time in REM sleep and a toddler spends approximately 30% of their time in REM sleep. From childhood through to adulthood, individuals spend approximately 20% of their time in REM sleep. (1 mark)

b 1 mark each for any two functions of REM; 1 mark for any two functions of NREM:
- REM sleep appears to be necessary for effective brain function.
- REM sleep facilitates the formation of synaptic links.
- REM sleep aids in the recovery of brain tissue after trauma or damage.
- REM sleep appears to be necessary for the consolidation of long-term memory.
- NREM sleep appears to be essential for physical growth and repair of the body.
- NREM sleep helps the body recuperate from high levels of physical activity.
- NREM sleep helps in recovery from illness.
- 'Slow wave' NREM sleep allows the glymphatic system to remove accumulated waste products from our brain.

35 a The suprachiasmatic nucleus responds to differing amounts and intensities of light, promoting the release of melatonin to make us go to sleep when it is dark and decreasing melatonin to enable us to wake up when it is light. (1 mark)

b Melatonin is a hormone that induces sleep when its levels rise. (1 mark)

c GABA enables the mind and body to relax, so we can fall asleep and sleep soundly throughout the night. (1 mark)

Importance of sleep to mental wellbeing

Multiple-choice answers

1 B

Mild sleep deprivation is most likely to result in an increase in irritability.
The other options are incorrect because they would decrease due to sleep deprivation.

2 D

Studies have shown that participants with partial sleep deprivation are mainly deprived of REM sleep, resulting in difficulty concentrating and performing simple tasks.

A is incorrect because hallucinations and delusions may occur after long periods of significant/total sleep deprivation. B is incorrect because sleep deprivation, especially of REM, has a detrimental effect on memory. C is incorrect because individuals with REM sleep deprivation will experience a subsequent increase in the amount of REM sleep on later nights.

3 C

The detrimental effects of one night of full sleep deprivation is more applicable to the ability to perform simple tasks than being able to execute more complex tasks successfully, thereby making C correct and A incorrect.

B is incorrect because one night of full sleep deprivation does affect coordination. D is incorrect because the side effects would subside after a few nights of proper sleep.

4 A

Studies have shown that the affective and cognitive effects of 24 hours (one night) of full sleep deprivation are equivalent to a blood alcohol concentration (BAC) of 0.10.

B is incorrect because cognition would be impaired for both individuals. **C** is incorrect because both individuals would have difficulty concentrating. **D** is incorrect because both individuals are likely to be more uninhibited as a result of their state.

5 A

A is the only option that describes a behavioural response. The other options refer to cognitive effects of partial sleep deprivation.

6 A

Partial sleep deprivation affects cognitive functioning.

B is incorrect because emotional changes would not explain her poor memory. **C** is incorrect because music festivals do not exacerbate the effect of sleep deprivation. **D** is incorrect because hallucinations are sometimes associated with prolonged total sleep deprivation, not partial deprivation as in this scenario.

7 C

A full night's sleep deprivation is equivalent to a blood alcohol concentration (BAC) of 0.10, but as the scenario specified the effect on Phoenix's concentration, **C** is the best option.

While **A** had the correct BAC, the physiological response was not appropriate for the prompt. **B** and **D** are incorrect because the level of sleep deprivation required to be equivalent to BAC of 0.05% would occur after approximately 20 hours (not a full day).

8 C

Working outdoors exposes us to daylight, which entrains and preserves our circadian biological rhythms. All of the other options would lead to disruptions of circadian rhythms.

9 D

Delayed Sleep Phase Syndrome is typically caused by environmental factors and delayed release of the hormone melatonin.

A and **B** are incorrect because this makes a person want to go to sleep earlier. **C** is incorrect because exposure to sunlight entrains and preserves our sleep–wake cycle.

10 A

People with Advanced Sleep Phase Disorder (ASPD) often complain of sleep-maintenance insomnia because of their early morning awakening (between 2 a.m. and 5 a.m.) and inability to fall back asleep.

B is incorrect because this pertains to narcolepsy. **C** is incorrect because this pertains to hypersomnia. **D** is incorrect because evening activities are restricted by their need to sleep at a much earlier time than others.

11 D

In comparison to those who work during normal business hours, shift workers get less, and poorer-quality, sleep.

A and **C** are incorrect because shift workers usually find it hard to get enough sleep because of the disruption to their sleep–wake cycle. **B** is incorrect because sleep quality is adversely affected by trying to sleep at a different time to the norm.

12 A

As the natural human cycle is slightly longer than the 24-hour day (approximately 25 hours), workers adjust quicker if shifts are rotated to a later shift, as they can cope better with an extended day.

B is incorrect because lack of routine by constant change can disrupt the circadian rhythm. **C** is incorrect because this effectively shortens the day, which causes biological rhythms to be out of phase with exogenous zeitgebers. **D** is incorrect because random changes, especially moving to an earlier shift, disrupts biological rhythms.

13 B

Circadian rhythms have been shown to be influenced by our eating and drinking patterns.
A is incorrect because zeitgebers are environmental cues (not internal mechanisms) that regulate and entrain circadian rhythms. **C** is incorrect because adrenaline affects arousal in response to certain stimuli and does not regulate circadian rhythms. **D** is incorrect because exposure to blue light from technology only becomes an issue in the evening.

14 D

Light is the strongest entraining agent for synchronising circadian rhythms, as the SCN responds to differing intensities of light and subsequently regulates the activity of the pineal gland in the secretion of melatonin to control the sleep–wake cycle.
None of the other options are affected by differing levels of light.

Short-answer sample responses

15 a Any three of the following (3 marks):
- Lethargy
- Irritability
- Headaches
- Loss of concentration
- Inattention
- Difficulty completing low-level boring tasks
- Inefficiency
- Confusion
- Misperception
- When allowed to sleep, the subject would fall asleep more quickly and sleep longer than usual

b Any two of the following (2 marks):
- More pronounced inefficiency and fatigue
- Disorientation
- Visual hallucinations
- Psychotic behaviour including paranoia and delusions

16 a Research by Lamond and Dawson has consistently shown that when participants are fully deprived of one night's sleep, they perform as badly in mental and physical tasks as when they are twice the legal blood alcohol limit for drivers in most Australian states. (1 mark) Volunteers who stayed awake for 20 hours produced performance levels equivalent to when they had a blood alcohol content of 0.05%. (1 mark) In those who stayed awake for about 24 hours, the performance level was equivalent to when they had a blood alcohol content of 0.10%. (1 mark)

b Performance on complex, intellectual tasks is not impaired by sleep deprivation, whereas an individual's ability to follow simple routines, as well as vigilance, attentiveness and mood, can be affected. (1 mark) Low-level boring tasks are the most likely to be affected, as sleep deprivation may affect motivation rather than ability. (1 mark)

17 a A circadian phase disorder occurs when gradual and sudden environmental changes disrupt our biological rhythms, putting us out of synchronicity with the Sun's cycle and causing changes to the sleep–wake cycle. (1 mark)

b In Advanced Sleep Phase Disorder, a person has difficulty staying awake unless they go to bed very early (usually between 6 p.m. and 9 p.m.), and often wakes up in the early morning (between 2 a.m. and 5 a.m.) (1 mark), whereas in Delayed Sleep Phase Syndrome, a person's sleep is delayed by 2 or more hours beyond what is considered a typical sleep schedule, going to sleep much later and wanting to sleep in the next morning (1 mark).

    c Night-shift workers often drive home in morning daylight, making it harder to reset their biological clocks and get little (or poor-quality) sleep during the middle of the day because human biological clocks promote sleepiness in the early morning hours. (1 mark)

    d Appropriately timed bright light exposure and avoidance of light at the wrong time of day (1 mark) have been shown to be effective strategies to accelerate entrainment for re-synchronising circadian rhythms (by changing the timing of melatonin release) to shift an individual's sleep–wake cycle to a desired, more appropriate schedule (1 mark).

18 ©VCAA 2015 MARKING GUIDELINES SB Q6a (ADAPTED)

    a As Ernie's sleep patterns would be disrupted due to his shift work, he would experience the effects of partial sleep deprivation. (1 mark) REM rebound occurs after partial sleep deprivation, with an increase in time spent in REM sleep in order to catch up. (1 mark) As dreaming occurs during REM sleep, REM rebound would result in an increase in (vivid/memorable) dreams. (1 mark)

> An alternative answer focusing on the acquisition of new skills was also acceptable: Ernie would need to consolidate and store new information he has learned at his new job. (1 mark) This would lead to an increase in REM sleep to enable consolidation of the new memories. (1 mark) As dreaming occurs during REM sleep, an increase in REM sleep results in an associated increase in Ernie's dreaming. (1 mark)

    b ©VCAA ©VCAA 2015 MARKING GUIDELINES SB Q6c (ADAPTED)

    Possible answers include:

- drowsiness or fatigue, microsleeps or droopy eyelids – resulting in errors in operating machinery
- decreased levels of arousal – cannot respond quickly or move out of the way of danger in the factory setting
- slower reflexes – cannot respond quickly or operate machinery effectively
- poor hand-eye coordination/hand tremors – cannot operate machinery effectively
- blurred vision/difficulty focusing eyes – may have difficulty seeing, increasing the possibility of error and impairing operation of machinery
- headaches – may cause difficulty concentrating and errors in operating machinery
- increased blood pressure – increased likelihood of cardiac event while operating machinery.

> 1 mark for identifying one physiological effect; 1 mark for explaining how this could lead to a workplace accident
>
> The following psychological effects were not appropriate to address the given prompt: cognitive impairment, decreased alertness/poor concentration and memory impairment.

19 a Zeitgebers are environmental cues (1 mark) that regulate and entrain circadian cycles to synchronise the internal body clock with the world around us (1 mark).

    b For each of the following, 1 mark for describing how they act as zeitgebers; 1 mark for explaining how they fit in to a healthy sleep–wake cycle.

        i People's circadian rhythms or cycles are influenced by the amount and intensity of light to which they are exposed. Exposure to daylight (or appropriately timed bright light) entrains a proper circadian rhythm which boosts wakefulness during the day and promotes sleep at an appropriate time at night.

        ii The blue light from technology such as from televisions and computer screens acts like sunlight, lowering your melatonin levels and making it harder to fall asleep. A healthy sleep routine would limit the use of such devices at night in the lead-up to bed and keep them out of the bedroom.

        iii Fluctuations in ambient temperature occur in connection with the day–night cycle: warmer in the day when we should be active and alert, and colder at night when we should be sleeping. Falling asleep is easier if the ambient temperature of our sleep environment is at a cool but comfortable level, ideally 16–18°C.

iv Daily routines, such as regular mealtimes, can help entrain and maintain our circadian rhythms. Alterations to the timing of food intake have the potential to cause misalignment of metabolic processes, making you feel alert or tired at different times than those you've become accustomed to. In order to maintain an appropriate sleep–wake cycle, we should endeavour to preserve a regular meal schedule and avoid stimulants for several hours before bedtime.

Chapter 4 Area of Study 2

Defining mental wellbeing

Multiple-choice answers

1 D

The biopsychosocial framework seeks to describe and explain how biological, psychological and social factors interact to influence a person's physical and mental health.

A and **C** are incorrect because physical factors are biological in nature. **B** is incorrect because the biopsychosocial framework also applies to physical health.

2 B

B represents a social factor that may not be evident during an interview, and therefore least likely to be detected.

C is not appropriate to the prompt as a clinician applying a purely medical approach would investigate possible underlying physiological conditions. **A** and **D** are behaviours associated with certain mental disorders and would certainly be noticed regardless of which approach was used.

3 A

This option indicates that a mentally healthy person can function, maintain relationships with others and cope with the normal stresses of life.

B is incorrect because the options under functioning and emotional wellbeing should be swapped. **C** is incorrect because respecting cultural identity would fall under social wellbeing. **D** is incorrect because the first option combines aspects of social and emotional wellbeing, not a level of functioning.

4 D

Glen would be considered mentally healthy because he worked towards goals in the face of stressors and disappointments in his life.

A is incorrect because Glen did not avoid his problems. **B** is incorrect because Glen was able to focus on the needs of his family. **C** is incorrect because such support is not a criterion for mental health; it can also be given to those with a mental illness.

5 A

B is incorrect because it does not include a psychological strategy. **C** and **D** are incorrect because these options do not have a biological component.

6 A

Kumi's behaviour would be indicative of a mental health problem due to its interference with normal functioning.

B is incorrect because there is no suggestion of delusions or hallucinations. **C** is incorrect because the behaviours are not occurring in a social context. **D** is incorrect because there is nothing in the scenario to indicate a physiological cause for her condition.

7 C

Although it could be argued that brain damage may contribute to some disorders, it is not present in a large number of conditions, and so therefore would not be one of the criteria that psychologists would use to help them determine whether someone is suffering from a disorder.

All of the other options would be considered as part of the diagnostic process.

8 C

The only appropriate option here is a level of functioning well below the norm.

The other options can occur without the person concerned having a mental disorder. **A** is too vague – everyone has problems to deal with. **B** is incorrect because challenging the status quo is essential for societal change and development. **D** is incorrect, as this can occur just because they are different.

9 A

B is incorrect because external factors do have an impact on mental wellbeing. **C** and **D** are incorrect because there is no evidence to suggest that certain types of factors have more significant impact on mental health than others. **C** is also incorrect because the level of resilience is not necessarily governed by the number of setbacks, but more likely shaped by the type of setbacks and the person's response to them. **D** is also incorrect because external factors can influence resilience.

10 D

Although **A** and **D** have strategies that fit under the appropriate headings, **D** is better, as proper nutrition would improve resilience (as required by the prompt) whereas medication (**A**) would not be seen as the best initial option in this regard.

B is incorrect because having his friends come and joining a club are both social strategies. **C** is incorrect because genetic testing would not improve his resilience.

11 D

The concept of social and emotional wellbeing (SEWB) provides a multidimensional framework which acknowledges that factors from several domains may combine to influence mental wellbeing.

All of the other options are incorrect because this framework takes a holistic approach consistent with the biopsychosocial model, which looks at the interconnectedness of a variety of dynamics affecting the physical and mental health of Indigenous Australians. **A** is also incorrect because SEWB does not just focus on the individual but recognises the impact any issues would also have on the wellbeing of their family and community.

12 B

These symptoms described in the scenario **most** resemble those found in cases of anxiety disorders.

The defining symptom for Alzheimer's disease is deficits in memory, and while the symptoms described could occur for some patients as a part of the advanced progression of this disease, **A** would not apply in every case and so would not be an appropriate choice for the prompt. Even though Benita's physical symptoms are consistent with Parkinson's, the features of her apprehension are not, making **C** an incorrect choice. **D** would be incorrect because stress is a reaction to a particular situation and would not occur for 'no apparent reason'.

13 B

Anxiety can be distinguished from phobia because only anxiety can be helpful in mild amounts.

The other options are incorrect because they apply to both anxiety and phobias.

14 A

Anxiety disorders may result from insufficient levels of GABA.

B is incorrect because insufficient levels of dopamine are associated with depression and Parkinson's disease. **C** and **D** are incorrect because anxiety would occur if there were excessive levels of noradrenaline and glutamate.

15 B

A phobia is best defined as an irrational fear of a specific object or situation.

A is incorrect because phobias are irrational. **C** is incorrect because phobias are not always linked to illness. **D** is incorrect because this refers to an unconditioned response, whereas a phobia is a conditioned fear response.

Short-answer sample responses

16 a Mental health is a state of emotional and social wellbeing (1 mark) in which a person can fulfil their abilities (1 mark), cope with normal stresses of life (1 mark), work productively and be able to contribute to the community (1 mark).

b Mental health problems are common mental health complaints (1 mark) that are usually understandable reactions to personal and social problems (1 mark) and are generally not too severe or long lasting (1 mark).

c Mental illness describes a psychological dysfunction (1 mark) that is so disturbing as to cause distress to the person and others (1 mark), usually involving impairment in the ability to cope with everyday life (1 mark) and which persists for a significant (relatively long) period of time (1 mark).

d Resilience is a person's capacity to cope and deal constructively with change or challenges (1 mark) and bounce back to maintain or re-establish their social and emotional wellbeing in the face of difficult events (1 mark).

e Social and emotional wellbeing (SEWB) is the term preferred by Aboriginal and Torres Strait Islander peoples to express their holistic and multidimensional concept of health and the factors that influence it (1 mark). It expresses the idea that the SEWB of an individual is affected by the strength of their connections to seven domains: 1. body; 2. mind and emotions; 3. family and kinship; 4. community; 5. culture; 6. Country; and 7. spirituality and Ancestors (1 mark). SEWB also includes the influence of broader contexts including social, historical, political and cultural contexts (1 mark).

17 a Any three of the following (3 marks):

- The person suffers from distress and/or impaired functioning serious enough to warrant professional treatment.
- The person's behaviour is maladaptive or harmful to that person or others.
- The person's behaviour does not represent a deliberate, voluntary decision to behave in a certain way.
- The source of the distress lies within the person, not the environment (such as prejudice, poverty or other social forces that may lead a person to behave contrary to social norms).
- The person's behaviour violates culturally determined standards or acceptability.

b The difference between 'normal' and 'abnormal' is largely arbitrary. This is due in part to the fact that many symptoms associated with mental disorders differ from common psychological experiences, not in kind but only in degree. (1 mark) Another difficulty is that several disorders could share some of the same symptoms, which could lead to misdiagnosis. (1 mark)

18 Any three of the following (3 marks):

- Toby had been acting out of character for 6 months, which is more serious or prolonged than would be expected of grief.
- Toby asking his parents to do his shopping for him, not being able to go to work and wearing the same clothes demonstrates an inability to function and cope independently.
- Toby's lack of response to friends' text messages or telephone calls indicates an inability to maintain social relationships.
- Toby exhibits changes in his attitudes and an apathetic lack of interest, as he hasn't left the house for several weeks, has not showered for days, wears the same clothes, doesn't go to work, no longer exercises and is not answering to his friends.
- Toby's thoughts/feelings/behaviours are atypical because the scenario implies that he would normally shower regularly, change his clothes, respond to text messages or phone calls and leave the house for work or social reasons.

> Responses needed to clearly indicate three independent reasons why Toby's behaviours would be seen as abnormal by his psychologist, not merely repeat three observations from the scenario.

19 a As Eliana's depressed mood and her inability to maintain social relationships or function by going to school has been occurring for a significant amount of time (1 mark), she would most likely be at point **D** (mental illness) (1 mark).

  b Chen would most likely be at point **B** (mental health problem) (1 mark) due to elevated stress from multiple stressors. This condition should be temporary, resolving once he passes his driving test. (1 mark) (There is nothing to indicate any further issues or impact on his life that would place him lower on the continuum.)

  c Stefan would probably be at point **A** (1 mark) as his behaviour is indicative of mental health. He is in a positive state of mental wellbeing but does have stressors to deal with (1 mark).

  d Voula would probably be at point **C** (severe mental health problem) (1 mark) due to high levels of stress and low levels of coping, which have continued for some time (1 mark). This condition should be temporary and be resolved once she makes her decision to defer, thereby removing one of the main sources of stress. (Her sleep problems have not persisted long enough to indicate a lower level on the continuum.)

20 Any three of the following (marks allocated for inclusion of contrasting details within each row) (3 marks):

Stress and anxiety	Phobia
Would be considered as a 'normal' reaction in certain circumstances	Always seen as 'abnormal'
Moderate levels can be adaptive and useful	Always a maladaptive response
Can be triggered by a wide range of stimuli	Triggered by a specific stimulus
Avoidance of certain things could occur, but this is not necessarily the case	Avoidance of the phobic stimulus is a central behavioural symptom
Could have a detrimental effect on a person's functioning if not handled effectively	Drastically affects a person's functioning
Risk factors that could lead to the development of a mental health disorder	A clinically recognised mental disorder

> Occasionally, VCAA may format the answer as a blank table, in which case you would clearly contrast each concept within each row. Otherwise, each part of your answer should be clearly distinguished within sentences such as: 'Stress and anxiety are ..., whereas phobias are ...'.

Application of a biopsychosocial approach to explain specific phobia

Multiple-choice answers

1 A

> A phobia would trigger a maladaptive response to the stimulus.
> All the other options are common in both phobias and normal fear.

2 D

> People with phobias may have a deficiency of the inhibitory neurotransmitter GABA.
> **A** and **C** are incorrect because GABA is an inhibitory neurotransmitter, not an excitatory neurotransmitter.
> **B** is incorrect because GABA is not in excess in those with phobias.

3 B

In Mae's case, long-term potentiation describes the lasting strengthening of synaptic connections within neural pathways representing the association between enclosed spaces and her fear of them, making it more likely that she will react in a fearful way to the specific stimulus.

A and **D** are incorrect because they describe the effects of long-term potentiation, not the process itself. **C** is incorrect because the neural signals do not fire continuously, only when she encounters the phobic stimulus.

4 D

Behavioural models explain phobias as being precipitated (developing) through classical conditioning of a fear response and perpetuated (maintained) by operant conditioning (specifically negative reinforcement) of avoidance (escape) behaviour.

A is incorrect because the fear response is not positively reinforced. **B** is incorrect because observational learning is a cognitive model, not behavioural. **B** and **C** are incorrect because avoidance involves negative reinforcement.

5 C

Arachnophobia develops through the association of a conditioned stimulus (spiders) with an experience that caused intense fear.

A and **D** are incorrect because if this were the case, arachnophobia would be a reflex, not a conditioned response. **B** is incorrect because this explains how phobias are maintained by operant conditioning.

6 B

In the given scenario, the conditioned response is Djamila feeling anxious when she walks by the alley.

A is incorrect because the alley is the originally neutral, now conditioned stimulus. **C** is incorrect because this is the unconditioned stimulus. **D** is incorrect because this is the unconditioned response.

7 C

Catastrophic thinking occurs when an individual only thinks about irrational worst-case outcomes of an experience with the phobic stimulus.

A is incorrect because Kaito's phobia developed as a result of vicarious classical conditioning. **B** is incorrect because Kaito has not actually encountered a shark to form such a memory. **D** is incorrect because this refers to watching the movie *Deep Blue Sea* as being the experience that led to the development of his phobia.

8 B

While people with a mental illness may perceive themselves as different from others, this does not necessarily mean it is due to the effects of social stigma or prejudicial attitudes from others.

The other options are all social risk factors that would increase the likelihood of the development and progression of a mental illness.

9 D

Benzodiazepines imitate the activity of the neurotransmitter gamma-aminobutyric acid (GABA) by inhibiting postsynaptic neurons in the brain to calm the body and reduce arousal.

A is incorrect because postsynaptic neurons need to be inhibited to reduce arousal. **B** and **C** are incorrect because the focus of the treatment is the effect on postsynaptic neurons, not on presynaptic neurons.

10 A

Breathing retraining is an example of a biological approach to reduce physiological arousal.

B is incorrect because this is more indicative of mindfulness meditation. **C** and **D** are incorrect because breathing retraining focuses on physiological processes in order to reduce arousal.

11 D

Exposure therapy for phobias applies counterconditioning techniques to extinguish the classically conditioned fear response.

The other component of behaviourist models involves the avoidance response, which is perpetuated by operant conditioning, specifically negative reinforcement, which requires a different extinction process (making **A** incorrect). Knowledge of the relative influences of each form of conditioning would eliminate **B** and **C** because of the mismatch between the type of conditioning and the response it influences.

12 C

In systematic desensitisation to treat phobias, the fear that has become associated with a specific object or event is reduced through the processes of counterconditioning and extinction.

A and **D** are incorrect because these options pertain to the avoidance behaviour associated with the phobia itself. **B** is incorrect because this would only increase an aversive response to the phobic stimulus.

13 B

In systematic desensitisation, phobic stimuli are arranged in a hierarchy from least- to most-feared items.

A is incorrect because this pertains to cognitive behavioural therapy. **C** is incorrect because this describes the process of flooding. Because **A** and **C** are incorrect, **D** is incorrect by default.

14 D

Cognitive behavioural treatments include both verbal interventions and behavioural modification techniques.

A is incorrect because this provides information but not practice leading to behaviour change. While this may form part of a cognitive behavioural approach, **B** is incorrect because this method focuses specifically on behaviour change in relation to phobias, not on the cognitive aspects. **C** is incorrect because this may clear the mind to facilitate logical thought but does not provide ways to change maladaptive behaviour patterns.

Short-answer sample responses

15 a A phobia is an intense, overwhelming, irrational fear of a specific object or situation. (1 mark)

  b When suffering from a phobia, exposure to the feared stimulus creates feelings of stress, apprehension and tension – all of which are symptoms associated with anxiety. (1 mark)

16 a People with specific phobia may not produce enough GABA or their brains may not process it normally. (1 mark) As a result, they may feel increased anxiety because of sustained firing of neurons causing overstimulation or overactivation of physical responses to fear/anxiety (or an inability to calm down). (1 mark)

  b Benzodiazepine agents are GABA agonists that mimic the inhibitory action of the neurotransmitter that is lacking for people with a phobia. (1 mark) Benzodiazepine agents are often used in combination with psychotherapy to calm the patient down/reduce physiological arousal to help them cope with their phobic anxiety. (1 mark) By reducing the symptoms of 'panic', use of benzodiazepine agents helps the patient to be able to focus and think more clearly in order for psychotherapy to have a chance of success. (1 mark)

  c Breathing retraining is a technique that enables those with anxiety to have more control over their condition by teaching correct breathing habits in order to slow down breathing and prevent hyperventilation. (1 mark) As well as helping to reduce anxiety, breathing training may also help people feel as if they have more control over their anxiety. (1 mark)

17 a Phobias are formed as a result of classical conditioning. (1 mark) A stimulus that normally does not elicit fear suddenly takes on this role after being associated with a fearful episode that an individual has experienced personally or vicariously through observational learning. (1 mark) Once acquired, phobias are maintained through operant conditioning. (1 mark) When a person begins to avoid (escape) the conditioned anxiety-producing stimulus, they experience a decrease in anxiety, which negatively reinforces their behaviour and prevents extinction of their fear. (1 mark)

b From the behavioural perspective, phobias are considered to be classically conditioned responses. (1 mark) A stimulus that can elicit an unconditioned fear response has been paired with another previously neutral stimulus, which has thereby become a conditioned stimulus capable of eliciting a conditioned fear response. (1 mark) According to the behaviourist, what can be learned can also be unlearned. If the conditioned stimulus is repeatedly presented without the unconditioned stimulus, extinction will occur. (1 mark)

> Rather than focus on describing the steps involved, this question asks you to explain the theoretical assumptions that underpin the therapy.

c Systematic desensitisation is a type of behavioural intervention in which a person is taught to relax while encountering increasingly threatening forms of the feared stimulus. (1 mark) Through graduated repeated exposure, the fear is extinguished, or unlearned. (1 mark)

d In systematic desensitisation treatment, the therapist teaches the client how to progressively and completely relax their body to decrease the physiological symptoms of anxiety. (1 mark) Next, the therapist and client first make up a fear hierarchy. The hierarchy lists stimuli that the client is likely to find frightening. The client then ranks the stimuli from least frightening to most frightening. (1 mark) Beginning with the least-frightening stimulus, the therapist would ask the patient to gradually work upwards through the hierarchy, one step at a time, progressing to the next step once they were able to relax and cope when exposed to each level of stimulus until the fear is gradually extinguished. (1 mark)

18 a Repeated activation of synaptic connections in the neural pathway strengthens the association between the phobic stimulus (conditioned stimulus) and the fear or anxiety response (conditioned response). (1 mark) This makes it more likely that the individual will react in a fearful way to the specific stimulus and decreases the likelihood that what has been learned will be forgotten. (1 mark)

b For someone with a phobia, cognitive bias represents a pattern of inaccurate thinking whereby their perception of the phobic stimulus will always be distorted, with the object appearing bigger and more threatening than it really is. (1 mark) Their interpretation will be supported by memory bias in that they will only remember negative or threatening experiences with the stimulus that will most likely be blown out of all proportion to what actually happened. (1 mark) Catastrophic thinking would then occur, as they would only think about irrational worst-case outcomes, overestimating the potential dangers or negative implications of their experience with the phobic stimulus. (1 mark)

c The cognitive component of CBT teaches sufferers how to change their thoughts (e.g. thinking rationally about whether there actually are any risks). (1 mark) The behavioural component of CBT teaches sufferers how to change their reactions to anxiety-provoking situations (e.g. to engage in deep breathing or other relaxation techniques). (1 mark)

19 a Traumatic events can suddenly trigger anxiety disorders in some individuals. (1 mark) For example, individuals who have been bitten by a dog may become conditioned to fear dogs and develop a specific phobia. (1 mark)

b 1 mark for using correct terminology; 1 mark for explaining the process:

Researchers believe that children can learn fears and phobias through vicarious classical conditioning just by observing their parent's or other family member's fearful reaction to an event.

> Specific phobias are a learned response, so while there is some suggestion that genetic factors play a role in certain anxiety disorders, such answers would be incorrect in this context.

c Because phobias are based on anxieties and fears that are quite personal, other people can find it difficult to empathise with the sufferer and become judgemental of their behaviour. (1 mark) This might result in stigma and discrimination, which can make it much harder for people to speak openly about what they're going through and seek help, thereby perpetuating the problem. (1 mark)

d Psychoeducation would provide the patient, their family and supporters with information about the symptoms and recovery patterns of the disorder in order to: (1 mark)
  - challenge unrealistic or anxious thoughts, and (1 mark)
  - focus on adaptive behavioural change rather than encouraging avoidance behaviours (1 mark).

Maintenance of mental wellbeing

Multiple-choice answers

1 D

Good health habits are protective factors against mental health disorders.
Protective behaviours promote physical and mental wellbeing. The other options refer to risk factors that could accumulate to trigger and maintain a mental health disorder.

2 C

Exercise, meditation and relaxation are important factors in stress management because they can alleviate the physiological arousal associated with stress.
A is incorrect because they do not increase the arousal associated with the fight–flight–freeze response. **B** is incorrect because a person may be very conscious of the stressful situation while employing these techniques. **D** is incorrect because these strategies do not remove the stressor.

3 A

Challenging maladaptive thinking is part of a cognitive behavioural strategy.
Although the other options are also protective factors, **B** and **C** are incorrect because these are sociocultural factors and **D** is incorrect because this is a biological factor.

4 B

Mindfulness meditation reduces psychological arousal by getting the individual to focus on something in the current moment.
A is incorrect because this is more indicative of concentrative meditation. **C** is incorrect because these types of brain waves are associated with deep NREM sleep. **D** is incorrect because mindfulness meditation does not seek to control bodily processes with the power of the mind.

Short-answer sample responses

5 a Protective factors are an individual's strengths, resilience, social supports and positive patterns of behaviour that prevent or reduce the likelihood or severity of conditions due to the risk factors present. (1 mark)

  b i A healthy diet is crucial to provide the nutrients and energy required to enable proper brain functioning. (1 mark)

   ii Hydration is necessary to enable proper blood circulation to the brain to supply oxygen, glucose and nutrients that are necessary for it to function properly. (1 mark)

   iii Good sleep patterns improve concentration and energy levels and assist in faster recovery from illness. (1 mark)

   iv As well as increasing physical fitness to effectively cope and deal with the physiological effects of stress, exercise can boost mood, concentration and alertness. (1 mark)

  c A poor diet, inadequate sleep and inactivity can be both causes and consequences of a variety of physical ailments (1 mark), and consequently represent precipitating and/or perpetuating risk factors for a number of mental disorders (1 mark).

6 a Cognitive behavioural strategies aim to alleviate symptoms by helping patients to look at all the possible interpretations of life events, identifying irrational negative thinking along with the possible helpful interpretations of a situation (1 mark), thereby learning a more positive approach to change their behaviour to be more constructive and productive in everyday situations (1 mark).

b Mindfulness meditation involves bringing awareness to be fully present in the current moment in order to centre our thoughts and avoid distractions about what has happened in the past or what might happen in the future. (1 mark)

7 a Examples could include revitalisation of traditional language; revitalisation of traditional cultural practices, such as traditional medicines and/or healing practices, songs, ceremonies, dance and art. (1 mark for each example)

b Description of how they contribute to maintaining SEWB can include any two of:
- reinforces connections to Country, culture, family and the community
- can help lessen feelings of isolation
- promotes a sense of pride in cultural heritage
- increases hope for the future
- provides people with supportive networks
- provides a sense of belonging at the family and community level.

8 a Biological risk factor: Substance abuse

Reason: Substance abuse would lead to the potential impairment of Zac's brain/cognitive functioning (as a precipitating or perpetuating factor).

Potential psychological risk factor: Stress

Reason: Stress could contribute to the development of mental illness through poor coping strategies and/or through the development of anxious/helpless/hopeless thought patterns.

Potential psychological risk factor: Impaired reasoning

Reason: Impaired reasoning could cause or maintain negative beliefs, irrational thought patterns and/or cognitive biases that are associated with mental illness.

> This question involved two 2-mark responses: 1 mark each was awarded for correct identification of an appropriate risk factor from the scenario; a further mark was given for an appropriate reason for how the identified risk factor could contribute to the development and/or progression of Zac's mental health disorder.

b Approach strategy, or emotion-focused strategy, or problem-focused strategy (1 mark)

c ©VCAA 2017 MARKING GUIDELINES SB Q6c (ADAPTED)

Possible responses would be:
- A recommendation might be that Zac join an interest group or club, which would be a protective factor against his growing isolation.
- A recommendation might be that Zac attempt to re-connect with friends and family, which would increase resilience through restoring links with social supports.
- A recommendation might be that Zac seeks help from employment and/or other social services to help him deal with his financial stressors.

> 1 mark for the recommendation of a social strategy; 1 mark for a reason as to how this would increase resilience

d ©VCAA 2017 MARKING GUIDELINES SB Q6d (ADAPTED)

Possible responses would be:
- Zac could feel embarrassed about seeing a psychologist because people may see this as a sign of weakness, which could prevent him from continuing his sessions.
- Due to society having certain views about males not needing to seek help for problems because they are tough and stoic in their outlook, Zac may feel ashamed about seeking treatment.

> 1 mark for identifying a possible source of stigma; 1 mark for a reason as to how this would prevent Zac from seeking help

Chapter 5 Area of Study 3

Investigation design

Multiple-choice answers

1 C

A research hypothesis is a testable statement phrasing the prediction regarding the outcome of a research study.

A is incorrect because a hypothesis is made before any research, so the actual outcome would not yet be known. **B** is incorrect because a hypothesis is a reasoned prediction based on observations and/or prior research and theory, not merely an educated guess. **D** is incorrect because this describes a null hypothesis and does not clarify how the variables will be implemented within the study.

2 D

The statement is an example of an operational hypothesis as it not only states a predicted outcome for the experiment but how it will be shown in practice.

A is not appropriate because a theory incorporates a general body of knowledge providing background for, but not necessarily making, a prediction. **B** is wrong because a null hypothesis would state that there would be no difference between the two groups. **C** is incorrect because research hypotheses use more general terms or concepts (in this case: 'That observational learning would improve performance').

3 A

B is incorrect because reaction time was the dependent variable. **C** is incorrect because cognitive performance was the dependent variable (DV) operationalised as reaction time. **B**, **C** and **D** are incorrect because the aim of the study was to see the effects in people at various stages of life – the amount of alcohol consumed and subsequent changes to the blood alcohol concentration (BAC) are part of the methodology to do so.

4 B

The graph demonstrates that the higher the BAC, the greater (slower) the reaction time.

While **A** may be true of those with higher BAC, it is incorrect because there is nothing in the graph to demonstrate that this has occurred. **C** is incorrect because reaction times get worse (slower) with age. **D** is incorrect because reaction times increase in both of these conditions.

5 D

In order for results to be valid and able to be generalised, a sample to be used in psychological research must be representative of the population from which it is drawn.

A is incorrect because the key is to have sufficient participants in the sample: you would not need a large sample if the population of interest was small. **B** is incorrect because the use of 'significant' here is ambiguous, especially because 'significance' is a term usually applied in the context of statistical analysis of the results. **C** is incorrect because such samples, while quick and easy to gather, are usually biased, making the results invalid.

6 B

A stratified sample ensures that the subset derived from the population represents relevant groups and/or participant characteristics in proportion to their numbers within the population.

A and **C** are incorrect because these methods do not ensure that relevant characteristics are proportional to the population. **D** is incorrect because allocation would occur after gathering the sample through participant selection.

7 C

Research participants are said to be randomly assigned when they each have an equal chance of being assigned to either the experimental or control group.

A is incorrect because this option describes random selection. **B** is incorrect because this just describes participant allocation, not how it should be random. **D** is incorrect because this describes the single-blind procedure, which could result from the allocation if required by the experiment.

8 D

Because the control group would not be exposed to the independent variable (IV), they should be given no alcohol at all.

A and **B** are incorrect because these would be experimental groups, as they are exposed to the IV (alcohol). **C** is incorrect because this would comprise the baseline measure and experimental results for the experimental groups.

9 D

Confounding variables do affect the DV and cloud the results, preventing valid conclusions from the research.

A is incorrect because these are extraneous and potentially confounding variables that have been identified before the research and kept constant to eliminate their effect on the results. **B** is incorrect because while extraneous variables could affect the results, there are many that don't, making this option less appropriate than **D**. **C** is incorrect because this is the variable the experimenters are assessing and wanting to affect the results.

10 B

In an experiment, it is essential to control for extraneous variables so that a valid conclusion can be made about the effect of the independent variable on the dependent variable.

A is incorrect because this may be the case even when all variables are controlled. **C** is incorrect because it is the IV that affects the DV. **D** is incorrect because it is not essential for the hypothesis to be supported in order for results to be generalised.

11 C

The researchers are trying to control the placebo effect. The participants believe that they are experiencing an experimental condition, whereas in fact this condition has a null effect in terms of the independent variable (caffeine level). This condition has been introduced to counter any effect that the participants' expectations may have on their sleep patterns.

Because it is only the participants who are unaware that this is a fake condition, **B** is incorrect because it is a single-blind, not a double-blind, procedure. **A** and **B** are incorrect because the researchers are aware of which participants are being given caffeine or not. **D** is incorrect because there are basically only two levels of the IV.

12 A

In a double-blind procedure, the participants don't know which group they are in, making **C** incorrect. To avoid the influence of any bias or unconscious indication of the real versus placebo condition by the person interacting with the participants or how the initial data is handled, the research assistant would not know, making **B** and **D** incorrect.

13 D

The researchers in this case have employed an independent-groups design as each group of participants was knowingly exposed to different experimental conditions.

A is incorrect because the researchers are aware of what form of instruction the participants have received. **B** is incorrect because the participants only undertake one form of learning. **C** is incorrect because there is no indication that participants were matched in terms of baseline skill levels before being allocated to one of the groups.

14 A

Measures of central tendency (mean, median, mode) and measures of variability (such as standard deviation) are all examples of descriptive statistics.
B is incorrect because while often used, these statistics are not required in every case. **C** and **D** are incorrect because the mean and standard deviation are quantitative measures on a ratio scale of measurement, not nominal or ordinal scales.

15 B

By receiving similar results from a different group of participants, the investigation has demonstrated high repeatability.
A is incorrect because the methodology was suitable for the purpose. **C** and **D** are incorrect because there may be a number of confounding participant variables.

16 C

The ethical principle of confidentiality is designed to protect the participants' rights to privacy by ensuring that researchers do not publicise any information that may identify any of the participants who were a part of the study. **A** and **D** relate to preventing coercion of participants to be involved in the study and **B** pertains to the guideline that the opportunity for feedback should be given to participants about the results/findings of the research.

17 B

Before they participate in a research study, potential participants must realise that their involvement in the study must be voluntary, and it is up to the researchers to emphasise this fact.
A is incorrect because the experimental participants, if human, are entitled to feedback about the results of the study. **C** is incorrect because deception by the experimenter may be permitted under certain circumstances where full knowledge of the experiment may affect the participants' behaviour, and hence the results of the study. **D** is incorrect because if a participant shows any sign of distress, the researcher is obligated to stop the experiment and tend to the welfare of the participant, whether they request it or not.

Short-answer sample responses

18 a Based on the results obtained from the sample, a conclusion is a statement pertaining to the specific study, including whether or not the hypothesis has been supported. (1 mark) Generalisations apply these conclusions beyond the sample to the wider population in settings outside the study. (1 mark)

b Any two of the following (2 marks):
- Random (and stratified) participant selection to ensure a sample that is representative of the population
- Random allocation to the experimental and control groups (there may also be the need for a baseline measure)
- Adequate operationalisation and control of the IV and DV
- Extraneous variables are controlled to eliminate any confounding variables
- Relevant and appropriate statistical methods are used to analyse the data, which would calculate the probability of whether they are due to chance (statistical significance)

19 a Because researchers are able to systematically manipulate the independent variable (caffeine levels) in a formal experiment, they are able to test causal links between the variables involved. (1 mark)

b The researchers employed stratified sampling to ensure their sample had the same proportions of age and gender as the general population. (1 mark)

c The study has used a double-blind procedure. (1 mark) Both the participants and the researchers were unaware as to which condition the participants were under. To prevent experimenter bias, the research assistant only informed the researchers which group had received the placebo after they had analysed the data. (1 mark)

- **d** The double-blind procedure attempts to control the experimenter effect. (1 mark) The experimenter effect is said to have occurred if the experimenter's personal characteristics, expectations or treatment of the data influence the results of the experiment. (1 mark)
- **e** The experimental design used in this study was an independent-groups design. (1 mark) This design tries to control for participant variables by randomly allocating participants such that it is equally likely that an individual could be in the experimental or control group. (1 mark)
- **f** Any one of the following (1 mark):
  - Quick and easy to do
  - No order effects
  - The experiment can all be done at once
  - There is less chance of participants dropping out
- **g** Any one of the following (1 mark):
  - There is less control over participant variables.
  - More participants are needed than for repeated measures.
- **h** Any two of the following (2 marks):
  - Participants were **not** asked to indicate how much caffeine they would normally consume
  - No control over any additional intake of caffeine
  - No baseline for memory or learning ability
- **i** The experimental group is Group 1, who were supplied unlabelled caffeinated beverages for their consumption over the two-week trial. (1 mark)
- **j** The experimental group is exposed to the experimental condition (the independent variable) whereas a control group is not. (1 mark)
- **k** The control group acts as a basis for comparison against the experimental group (or to compare the effect of the IV with the experimental group). (1 mark)
- **l** Group 2 was under a placebo condition. (1 mark) They were receiving a fake 'treatment'; that is, a condition that mimicked the 'treatment' but which would have not had the properties under investigation in this study (in this case, decaffeinated beverages). (1 mark) The purpose of such a condition is to try to counter any effects due to participants' beliefs and/or expectations. (1 mark)

20 a That one night of sleep deprivation (sustained wakefulness) will produce performance impairment on a range of neurobehavioural tasks that is comparable to the effects of alcohol intoxication.

> 1 mark for appropriate operationalisation of the independent variable; 1 mark for appropriate operationalisation of the dependent variable; 1 mark for indicating the predicted relationship between the variables

- **b** Levels of BAC in the alcohol intoxication condition (1 mark); number of hours without sleep in the sustained wakefulness condition (1 mark).
- **c** Speed and/or accuracy and/or increases in the duration of responses on a variety of neurobehavioural performance tasks (1 mark)
- **d** Non-random convenience sampling (1 mark)
- **e** A repeated-measures design (1 mark)
- **f** Either of the following (1 mark):
  - To minimise individual differences due to participant variables between conditions
  - To minimise (economise on) the number of participants
- **g** Either of the following (1 mark):
  - Possible carryover effects, where knowledge from a previous condition affects or influences performance on a later condition (either improve due to *practice* or worsen, due to *boredom*).
  - The time interval between repeated testing may lead to changes in the participants' characteristics (i.e. history effects) potentially confounding the effect of the treatment variable.

h *Counterbalancing* is a procedure used to try to balance order effects (1 mark):
- by placing half the participants into one experimental condition first, with the remainder in the other condition, then swapping conditions (1 mark)
- in the study by Lamond and Dawson, half would be sleep-deprived while the other half would be under the influence of alcohol, then swap conditions for the second part of the study (1 mark).

i This study found that 17–20 hours of sleep deprivation produced performance decrements equivalent to those observed at a BAC of 0.05 (1 mark) and 24–25 hours (one night) of full sleep deprivation produced performance impairment equivalent to that observed at a BAC of 0.10%, or twice the legal limit for driving, working and/or operating dangerous machinery (1 mark).

j The main hypothesis was supported. (1 mark) The researchers concluded that moderate levels of sustained wakefulness (>20 hours) produced a reduction in performance equivalent to that observed at moderate levels of alcohol intoxication in social drinkers. (1 mark)

k As an example, shift workers on a first-night shift greater than 24 hours in a roster would show a neurobehavioural performance decrement similar to or greater than the legal BAC limit for driving, working and/or operating dangerous machinery. (1 mark)

l
- Only 22 participants were used, which is too small a sample size to infer significant results. A sample size of 100 would have been ideal.
- Participants were aged between 19 and 26 years, which is not representative of the general age of the adult population.
- The gender of the participants was not specified. Gender may have been a confounding variable as sleep deprivation and/or alcohol consumption may have affected males and females differently. In addition, if the gender of participants did not represent a stratified sample of the population; it may not have been a representative sample.
- Participants were recruited via advertisements placed around universities, which was likely to result in a sample with specific age, income and education levels that were not representative of the general population.
- It is hard to generalise results in a laboratory setting to real-life settings, as these environments usually bear little resemblance to actual tasks and settings.
- Not all functions involved in real-life tasks, such as driving, were used and assessed.

> This question involved two 2-mark responses: 1 mark each was awarded for identification of a limitation of the research design and methodology; a further mark was given for an appropriate reason for why it was a limitation.

m Any three of the following (3 marks):
- Informed consent: Researchers needed to explain all of the necessary details regarding the study in order for the participants to reach an informed decision to agree to be a part of the experiment. Such information should point out any potential risks that may be present in the research design.
- Voluntary participation: Participants became a part of the experiment because they chose to do so.
- Withdrawal rights: Participants also retained the right to remove themselves from the research situation at any point if they chose to do so.
- Confidentiality: Researchers needed to ensure that only relevant demographic data was collected and reported in such a way that the participants' identities were not revealed.

- Non-maleficence: Researchers had to ensure the safety and wellbeing of the participants, especially as some of the experimental conditions would be putting them into physiological states that would increase the likelihood of accidental injury. The researchers also had to monitor the participants for any adverse physiological reactions to any of the experimental conditions and remove them for immediate medical attention should any such reaction or injury occur.

21 Grade 4 children in Victoria who watched the literacy program on television will show a greater increase in the score between literacy tests A and B than participants who watched cartoons of their choice.

1 mark for statement of the population, the independent variable and dependent variable; 1 mark for appropriate description of the direction of change in the dependent variable

22 ©VCAA 2009 E2 MARKING GUIDELINES SB Q11 (ADAPTED)

Independent variable: watching eiher the literacy program or watching cartoons of choice (1 mark)

Dependent variable: literacy skills in children (operationalised as the difference/improvement in score between literacy tests A and B) (1 mark)

23 ©VCAA 2009 E2 MARKING GUIDELINES SB Q12 (ADAPTED)

Participants were randomly allocated as the computer placed one of each pair, by chance, into each of the groups. (1 mark) Each participant had an equal chance of being in either group. (1 mark)

24 ©VCAA 2009 E2 MARKING GUIDELINES SB Q13 (ADAPTED)

a
- Matched participants (matched pairs) or matched subjects
- Participants were paired based on similarity of relevant characteristics (literacy skills [score on test A] and gender). One participant from each pair was then allocated to either the experimental group or the control group.

1 mark for identifying the experimental design used in this case; 1 mark for explaining the answer

b Independent groups:

Disadvantages
- More participants are needed to obtain the same strength of results
- Extraneous variables due to participant characteristics (such as gender and literacy skills) are not controlled

Repeated measures:

Disadvantages
- Order effects (such as practice, learning or boredom) may be a confounding variable and interfere with results
- More time-consuming because an extra four weeks would be required

1 mark for identifying another experimental design; 1 mark for identifying a disadvantage of that design in this case

25
- Children volunteered by parents – not a random sample
- Participant characteristics (e.g. Level of motivation? Need for extra help?)
- Extra help given by teachers, especially if they knew who was watching the literacy program
- Tests administered by Grade 4 teachers – not standardised, open to alteration/bias and so on

1 mark for identifying one uncontrolled variable; 1 mark for explaining how this variable would have affected the results

26 ©VCAA 2009 E2 MARKING GUIDELINES SB Q17 (ADAPTED)

These results should not be generalised. (1 mark)

Any one of the following: (1 mark)
- Participants were not selected randomly
- Only participants who volunteered in the first 100 were selected
- Not every Grade 4 child had an equal chance of being selected

27 ©VCAA 2009 E2 MARKING GUIDELINES SB Q18 (ADAPTED)

Any two of the following (2 marks):
- A thorough explanation of the study's findings
- The right to have their data removed after the experiment
- Details about where and how to obtain psychological help (i.e. counselling) if required